T0291868

CAMBRIDGE LIBRARY COLLECTION

Books of enduring scholarly value

Darwin

Two hundred years after his birth and 150 years after the publication of 'On the Origin of Species', Charles Darwin and his theories are still the focus of worldwide attention. This series offers not only works by Darwin, but also the writings of his mentors in Cambridge and elsewhere, and a survey of the impassioned scientific, philosophical and theological debates sparked by his 'dangerous idea'.

Vestiges of the Natural History of Creation

Vestiges of the Natural History of Creation was published anonymously in 1844. Starting with the genesis of the solar system, it progresses systematically through such topics as the formation of the earth, the origins of marine life, the emergence of reptiles, birds and other life forms, and the evolution of human life. Drawing widely upon contemporary ideas from astronomy, biology, geology, linguistics, and anthropology, it seeks to establish the 'hypothesis of an organic creation by natural law'. Preceding Darwin's Origin of Species by fifteen years, Vestiges ignited a storm of controversy by pitting natural law and its role in what would soon come to be known as 'evolution' against the generally accepted Victorian belief that the universe was created by God. In 1884, it was revealed that the author was the British publisher Robert Chambers. This fifth edition of the work also contains its 1845 sequel, Explanations.

Cambridge University Press has long been a pioneer in the reissuing of out-of-print titles from its own backlist, producing digital reprints of books that are still sought after by scholars and students but could not be reprinted economically using traditional technology. The Cambridge Library Collection extends this activity to a wider range of books which are still of importance to researchers and professionals, either for the source material they contain, or as landmarks in the history of their academic discipline.

Drawing from the world-renowned collections in the Cambridge University Library, and guided by the advice of experts in each subject area, Cambridge University Press is using state-of-the-art scanning machines in its own Printing House to capture the content of each book selected for inclusion. The files are processed to give a consistently clear, crisp image, and the books finished to the high quality standard for which the Press is recognised around the world. The latest print-on-demand technology ensures that the books will remain available indefinitely, and that orders for single or multiple copies can quickly be supplied.

The Cambridge Library Collection will bring back to life books of enduring scholarly value (including out-of-copyright works originally issued by other publishers) across a wide range of disciplines in the humanities and social sciences and in science and technology.

Vestiges of the Natural History of Creation

Together with Explanations: A Sequel

Robert Chambers

CAMBRIDGE UNIVERSITY PRESS

Cambridge, New York, Melbourne, Madrid, Cape Town, Singapore,
São Paolo, Delhi, Dubai, Tokyo

Published in the United States of America by Cambridge University Press, New York

www.cambridge.org
Information on this title: www.cambridge.org/9781108001670

This edition first published 1846
This digitally printed version 2009

ISBN 978-1-108-00167-0 Paperback

VESTIGES

OF

THE NATURAL HISTORY

OF

CREATION.

Fifth Edition.

LONDON:
JOHN CHURCHILL, PRINCES STREET, SOHO.
MDCCCXLVI.

CONTENTS.

PAGE

THE BODIES OF SPACE—THEIR ARRANGEMENTS AND FORMATION 1

CONSTITUENT MATERIALS OF THE EARTH, AND OF THE OTHER BODIES OF SPACE 29

THE EARTH FORMED—ERA OF THE PRIMARY ROCKS . 46

COMMENCEMENT OF ORGANIC LIFE — SEA PLANTS, CORALS, ETC. 55

ERA OF THE OLD RED SANDSTONE—FISHES ABUNDANT 68

SECONDARY ROCKS—ERA OF THE CARBONIFEROUS FORMATION—COMMENCEMENT OF LAND PLANTS . . . 79

ERA OF THE NEW RED SANDSTONE—TERRESTRIAL ZOOLOGY COMMENCES WITH REPTILES—FIRST TRACES OF BIRDS 97

ERA OF THE OOLITE—COMMENCEMENT OF MAMMALIA . 110

ERA OF THE CRETACEOUS FORMATION 123

ERA OF THE TERTIARY FORMATION — MAMMALIA ABUNDANT 131

PAGE

ERA OF THE SUPERFICIAL FORMATIONS—COMMENCEMENT OF PRESENT SPECIES. 141

GENERAL CONSIDERATIONS RESPECTING THE ORIGIN OF THE ANIMATED TRIBES 153

PARTICULAR CONSIDERATIONS RESPECTING THE ORIGIN OF THE ANIMATED TRIBES 173

HYPOTHESIS OF THE DEVELOPMENT OF THE VEGETABLE AND ANIMAL KINGDOMS 197

AFFINITIES AND GEOGRAPHICAL DISTRIBUTION OF ORGANISMS 240

EARLY HISTORY OF MANKIND 294

MENTAL CONSTITUTION OF ANIMALS 338

PURPOSE AND GENERAL CONDITION OF THE ANIMATED CREATION 373

NOTE CONCLUSORY 410

APPENDIX 415

THE BODIES OF SPACE,

THEIR ARRANGEMENTS AND FORMATION.

It is familiar knowledge that the earth which we inhabit is a globe of somewhat less than 8000 miles in diameter, being one of a series of eleven which revolve at different distances around the sun, and some of which have satellites in like manner revolving around them. The sun, planets, and satellites, with the less intelligible orbs termed comets, are comprehensively called the solar system, and if we take as the uttermost bounds of this system the orbit of Uranus (though the comets actually have a wider range), we shall find that it occupies a portion of space not less than three thousand six hundred millions of miles in diameter. The mind fails to form an exact notion of a portion of space so immense; but some faint idea of it may be obtained from the fact, that, if the swiftest race-horse ever known had begun to traverse it, at full speed, at the time of the birth of Moses, he would only as yet have accomplished half his journey.

It has long been concluded amongst astronomers, that the stars, though they only appear to our eyes as brilliant points, are all to be considered as suns, representing so many solar systems, each bearing a general resemblance to our own. The stars have a brilliancy and apparent magnitude which we may safely presume to be in proportion to their actual size and the distance at which they are placed from us. Attempts have been made to ascertain the distance of some of the stars by calculations founded on parallax, it being previously understood that, if a parallax of so much as one second, or the 3600th of a degree, could be ascertained in any one instance, the distance might be assumed in that instance as not less than 19,200,000 millions of miles! In the case of the most brilliant star, Sirius, even this minute parallax could not be found; from which of course it was to be inferred that the distance of that star is something beyond the vast distance which has been stated. In some others, on which the experiment has been tried, no sensible parallax could be detected; from which the same inference was to be made in their case. But a sensible parallax of about one second has been ascertained in the case of the double star, ά ά, of the constellation of the Centaur,* and one of the third of that amount for the double star, 61 Cygni; which gives reason to presume that the distance of the former may be about nineteen millions of millions of miles, and the latter of much greater amount. If we suppose that similar intervals exist between all the stars, we shall readily see that the space occupied by even the comparatively small number visible to the naked eye must be vast beyond all powers of conception.

The number visible to the eye is about three thousand; but when a telescope of small power is directed to the heavens, a great number more come into view, and the number is ever increased in proportion to the increased power of the instrument. In one place, where they are more thickly sown than elsewhere, Sir William Herschel reckoned that fifty thousand passed over a field of view two degrees in breadth in a single hour. It was first surmised by the ancient philosopher, Democritus, that the faintly white zone which

* By the late Mr. Henderson, Professor of Astronomy in the Edinburgh University, and Lieutenant Meadows.

spans the sky under the name of the Milky Way, might be only a dense collection of stars too remote to be distinguished. This conjecture has been verified by the instruments of modern astronomers, and some speculations of a most remarkable kind have been formed in connexion with it. By the joint labours of the two Herschels, the sky has been "gauged" in all directions by the telescope, so as to ascertain the conditions of different parts with respect to the frequency of stars. The result has been a conviction that, as the planets are parts of solar systems, so are solar systems parts of what may be called Astral Systems—that is, systems composed of a multitude of stars, bearing a certain relation to each other. The astral system to which we belong, is conceived to be of an oblong, flattish form, with a space wholly or comparatively vacant in the centre, while the extremity in one direction parts into two. The stars are most thickly sown in the outer parts of this vast ring, and these constitute the Milky Way. Our sun is believed to be placed in the southern portion of the ring, near its inner edge, so that we are presented with many more stars, and see the Milky Way much more clearly, in that direction, than towards the north, in which line our eye has to traverse the vacant central space. Nor is this all. A motion of our solar system with respect to the stars, first suggested by Sir William Herschel, in 1783, has since been verified by the exact calculations of M. Argelander, late director of the Observatory at Abo. The sun is proceeding towards a point in the constellation Hercules. It is, therefore, receding from the inner edge of the ring. Motions of this kind, through such vast regions of space, must be long in producing any change sensible to the inhabitants of our planet, and it is not easy to grasp their general character; but grounds have nevertheless been found for supposing that not only our sun, but the other suns of the system, pursue a wavy course round the ring, *from west to east*, crossing and recrossing the middle of the annular circle. "Some stars will depart more, others less, from either side of the circumference of equilibrium, according to the places in which they are situated, and according to the direction and the velocity with which they are put in motion. Our sun is probably one of those which depart furthest from it, and descend furthest into the empty space within the ring."* According to this view, a time

* Professor Mossotti, on the Constitution of the Sidereal System, of which the Sun forms a part.—*London, Edinburgh, and Dublin Philosophical Magazine*, February, 1843.

may come when we shall be much more in the thick of the stars of our astral system than we are now, and have of course much more brilliant nocturnal skies; but it may be countless ages before the eyes which are to see this added resplendence shall exist.

The evidence of the existence of other astral systems besides our own is much more decided than might be expected, when we consider that the nearest of them must needs be placed at a mighty interval beyond our own. The elder Herschel, directing his wonderful tube towards the *sides* of our system, where stars are planted most rarely, and raising the powers of the instrument to the required pitch, was enabled with awe-struck mind to see suspended in the vast empyrean astral systems, or, as he called them, firmaments, resembling our own. Like light cloudlets to a certain power of the telescope, they resolved themselves, under a greater power, into stars, though these generally seemed no larger than the finest particles of diamond dust.* The general forms of these systems (*astral nebulæ*) are various. So also are the distances, as proved by the different degrees of telescopic power necessary to bring them into view.

* See Appendix, A.

The furthest observed by the astronomer were estimated by him as thirty-five thousand times more remote than Sirius, supposing its distance to be about twenty millions of millions of miles. It would thus appear, that not only does gravitation keep our earth in its place in the solar system, and the solar system in its place in our astral system, but it also may be presumed to have the mightier duty of preserving a local arrangement between that astral system and an immensity of others, through which the imagination is left to wander on and on without limit or stay, save that which is given by its inability to grasp the unbounded.

The two Herschels have in succession made some other remarkable observations on the regions of space. They have found, within the limits of our astral system, and generally in its outer fields, a great number of nebulæ of a different character from the above; some of vast extent and irregular figure; others of shape more defined; others, again, in which small bright nuclei appear here and there over the surface. Between this last form and another class of objects, which appear as clusters of nuclei with nebulous matter around each nucleus, there is but a step in what appears a chain of related things. Next we have what are called *photospheres* or *nebulous stars,*—luminous spherical objects, bright in the centre and dull towards the extremities. These appear to be only an advanced condition of the class of objects above described. Finally, nebulous stars exist in every stage of concentration, down to that state in which we see only a common star with a slight *bur* around it. It may be presumed that all these are but stages in a progress, just as if, seeing a child, a boy, a youth, a middle-aged, and an old man together, we might presume that the whole were only variations of one being. Are we to suppose that we have got a glimpse of the process through which a sun goes between its original condition, as a mass of diffused nebulous matter, and its full-formed state as a compact body? We shall see how far such an idea is supported by other things known with regard to the occupants of space, and the laws of matter.

A superficial view of the astronomy of the solar system gives us only the idea of a vast luminous body (the sun) in the centre, and a few smaller, though various sized bodies, revolving at different distances around it; some of these, again, having smaller planets (satellites) revolving around them.

There are, however, some general features of the solar system which, when a profounder attention makes us acquainted with them, strike the mind very forcibly.

It is, in the first place, remarkable, that the planets all move nearly *in one plane*, corresponding with the centre of the sun's body. Next, it is not less remarkable, that the motion of the sun on its axis, those of the planets around the sun, and the satellites around their primaries,* and the motions of all on their axes, are *in one direction*—namely, from west to east. Had all these matters been left to accident, the chances against the uniformity which we find would have been, though calculable, inconceivably great. Laplace states

* There is an exception, but I believe apparent only, in the motion of the satellites of Uranus, which, compared with the rest, is retrograde. The axes of the planets are, as is well known, at various degrees of inclination to their orbits; for which there must have been a cause in the circumstances under which the planets were produced. The axis of Uranus is removed but eleven degrees from the plane of his orbit: I suggest, as the explanation of the apparent exception, that what we call the north pole of this planet is in reality the south, the axis having passed across the plane of the orbit, so that the planet may be said to be in that small measure upside down. It will be observed, that between the admitted and the suggested arrangement, there is only a difference of 22 degrees.

them at four millions of millions to one. It is thus powerfully impressed on us, that the uniformity of the motions, as well as their general adjustment to one plane, must have been a consequence of some cause acting throughout the whole system.

Some of the other relations of the bodies are not less remarkable. It is of little consequence that the larger planets are towards the outside of the system, since there is an absence of regularity in the gradation in this respect. In the series of comparative densities we find something approaching to a regular gradation: they stand thus in decimals, the Earth being considered as 1—Mercury, 2·95; Venus, ·99; Earth, 1; Mars, ·79; Jupiter, ·23; Saturn, ·11; Uranus, ·26; the last being the only very decided violation of the rule. Then the distances are curiously relative. It has been found that, if we place the following line of numbers,—

0	3	6	12	24	48	96	192,

and add 4 to each, we shall have a series denoting the respective distances of the planets from the sun. It will stand thus—

4	7	10	16	28	52	100	196
Merc.	Venus.	Earth.	Mars.		Jupiter.	Saturn.	Uranus.

It will be observed that the first row of figures goes on from the second on the left hand in a succession of duplications, or multiplications by 2. Surely there is here a most surprising proof of unity in the solar system. It was remarked, when this curious relation was first detected, that there was the want of a planet corresponding to 28; the difficulty was afterwards considered as overcome, by the discovery of four small planets revolving at nearly one mean distance from the sun, between Mars and Jupiter. The distances bear an equally interesting mathematical relation to the times of the revolutions round the sun. It has been found that, with respect to any two planets, the squares of the times of revolutions are to each other in the same proportion as the cubes of their mean distances,—a most surprising result, for the discovery of which the world was indebted to the illustrious Kepler. Sir John Herschel truly observes— "When we contemplate the constituents of the planetary system from the point of view which this relation affords us, it is no longer mere analogy which strikes us, no longer a general resemblance among them, as individuals independent of each other, and circulating about the sun, each according to its own peculiar nature, and connected with it by its own peculiar tie. The resemblance is now perceived to be a true *family likeness*; they are bound up in one chain—interwoven in one web of mutual relation and harmonious agreement, subjected to one pervading influence, which extends from the centre to the farthest limits of that great system, of which all of them, the Earth included, must henceforth be regarded as members."*

Connecting what has been observed of the series of nebulous stars with this wonderful relationship seen to exist among the constituents of our system, and further taking advantage of the light afforded by the ascertained laws of matter, modern astronomers have suggested the following hypothesis of the formation of that system.

Of nebulous matter in its original state we know too little to enable us to suggest how nuclei should be established in it. But, supposing that, from a peculiarity in its constitution, nuclei are formed, we know very well how, by virtue of the law of gravitation, the process of an aggregation of the neighbouring matter to those nuclei should proceed, until masses more or less solid should become detached from the rest. It is well known,

* Treatise on Astronomy.

that, when fluid matter collects towards or meets in a centre, there are many chances against its meeting so directly as to produce rest; most frequently, the various momenta are opposed in such an oblique way, as to cause a rotatory motion. This result is familiarized to us in the whirlwind and the whirlpool—nay, on so humble a scale as the water sinking through the aperture of a funnel. It thus becomes certain, that when we arrive at the stage of a nebulous star, there is a great probability of a rotation on an axis having been commenced.

Speculation on admitted laws of mechanical philosophy has followed out this idea to most remarkable effects. It is found that, the instant a mass begins to rotate, the outer parts tend to fly off: in other words, the law of centrifugal force begins to operate. There are, then, two forces acting in opposition to each other, the one attracting *to*, the other throwing *from*, the centre. While these in the above case remained exactly counterpoised, the mass would necessarily continue entire; but the least excess of the centrifugal over the attractive force would be attended with the effect of separating the outer parts from the mass. These outer parts would then be left as a ring round the central body, which ring would continue to revolve with the velocity possessed by the central mass at the moment of separation, but not necessarily participating in any changes afterwards undergone by that body. This is a process which would be repeated as soon as a new excess arose in the centrifugal over the attractive forces working in the parent mass, and this until the mass attained the ultimate limits of the condensation which its constitution imposed upon it.

If these rings consisted of matter nearly uniform throughout, they would probably continue each in its original form; but there are many chances against their being uniform in constitution. The unavoidable effects of irregularity in their constitution would be to cause them to gather towards centres of superior solidity, by which the annular form would, of course, be destroyed. The ring would, in short, break into several masses, the largest of which would be likely to attract the lesser into itself. The whole mass would then necessarily settle into a spherical form by virtue of the law of attraction; in short, would become a planet revolving round the sun. Its rotatory motion would, of course, continue, and satellites might then be thrown off in turn from its body in exactly the same way as the primary planets had been thrown off from the sun.

Such were the suggestions of the elder Herschel and Laplace as to our cosmogony; the latter of whom proved its "dynamical possibility" by well-known physical laws. In the time of these great men, it received only the further support of two or three features of the actual condition of the solar system, which appear at first in the light of exceptions to the ordinary forms. The rings which surround the body of Saturn appeared to Laplace as instances in which the breaking up and consolidation of the annular form of the planetary masses had not taken place: in those instances, it was to be presumed that the matter had been sufficiently equable to remain in that form which in all others was transient. It was equally admissible, theoretically, that, when a ring broke up, the fragments might spherify separately. To this idea the four little planets moving between Mars and Jupiter, at nearly one mean distance from the sun, answered exactly; the unusual and extremely various inclinations of their orbits to the plane of the ecliptic, being apparently the monument of an inequality of form and of movement in the ring, which had led to the fragments keeping apart. The existence of the Zodiacal Light, afterwards to be adverted to, might be considered as a further memorial of the process by which the solar nebula had passed into its present arrangements. Perhaps, however, the most tangible corroboration which the nebular cosmogony, as it is called, had received up to that time, was the well known oblately spheroidal form of the planets. This peculiar shape is an incontestable proof that the planets were at one time in a fluid state—the condition which the nebular cosmogony supposes.

The nebular cosmogony also obtains a remarkable support in what would at first seem to militate against it—the existence in our firmament of several thousands of solar systems, in which there are more than one sun. These are called double and triple stars. Some double stars, upon which careful observations have been made, are found to have a regular revolutionary motion round each other in ellipses. This kind of solar system has also been observed in what appears to be its rudimental state, for there are examples of nebulous stars containing two and three nuclei in near association. At a certain point in the confluence of the matter of these nebulous stars, they would all become involved in a common revolutionary motion, linked inextricably with each other, though it might be at sufficient distances to allow of each distinct centre having afterwards its attendant planets. We have seen that the law which causes rotation in the single solar masses, is exactly the same which produces the familiar phenomenon of a small whirlpool or dimple in the surface of a stream. Such dimples are not always single. Upon the face of a river where there are various contending currents, it may often be observed that two or more dimples are formed near each other with more or less regularity. These fantastic eddies, which the musing poet will sometimes watch abstractedly for an hour, little thinking of the law which produces and connects them, are an illustration of the wonders of binary and ternary solar systems.

Such was the position of this interesting hypothesis when Professor Plateau, of Ghent, advanced it by experiment to a point at which we may almost be said to see it passing into the region of ascertained truths. Before going further, it is necessary to remind the reader of a principle which every investigation of nature only serves to confirm, that, in natural phenomena, there is no distinction of great and small within themselves. A dew-drop and a planet spherify by similar laws. The oblately spheroidal form is assumed by a flexible hoop whirling on the table of a lecture-room, exactly as it is by a globe of thousands of miles diameter revolving on its majestic round through space. And astronomers determine that the rapidity of Mercury is greater than that of Saturn, precisely for the same reason that when we wheel a ball round by a string, and allow the string to wind round one of the fingers, the ball flies the quicker as the string is shortened. Keeping these things in view, and regarding nature with the fearless simplicity which never can be inconsistent with true reverence, we may be in some degree prepared to hear of the Ghent experiments,—which, divested of technical terms, were nearly as follows. Placing a mixture of water and alcohol in a glass box, and therein a small quantity of olive oil of density precisely equal to the mixture, we have in the latter *a liquid mass relieved from the operation of gravity*, and free to take the exterior form given by the forces which may act upon it. In point of fact, the oil, by virtue of the law of molecular attraction, instantly takes a globular form. A vertical axis being introduced through the box, with a small disc upon it, so arranged that its centre is coincident with the centre of the globe of oil, we turn the axis at a slow rate, and thus set the oil-sphere in rotation. "We then presently see the sphere *flatten at its poles* and *swell out at its equator*, and thus realize on a small scale an effect which is admitted to have taken place in the planets." The spherifying forces are of different natures, that of molecular attraction in the case of the oil, and of universal attraction in that of the planet; but the results are "analogous if not identical." Quickening the rotation makes the figures more oblately spheroidal. When it comes to be so quick as two or three turns in a second, "the liquid sphere first takes rapidly its maximum of flattening, then becomes hollow above and below around the axis of rotation, stretching out continually in a horizontal direction, and finally, abandoning the disc, is *transformed into a perfectly regular ring*." At first, this remains connected with the disc by a thin pellicle of oil; which, however, on the disc being stopped, breaks and disappears, and the ring then becomes completely disengaged. The only observable difference between this ring and that of Saturn, is that it is rounded, instead of being flattened; but this is accounted for by the learned professor in a satisfactory way.

A little after the stoppage of the rotatory motion of the disc, the ring of oil, losing its own motion, gathers once more into a sphere. If, however, a smaller disc be used, and its rotation continued after the separation of the ring, rotatory motion and centrifugal force will be generated in the alcoholic fluid, and the oil-ring, thus prevented from returning into the globular form, divides itself into *several isolated masses, each of which immediately takes the globular form.* These "are almost always seen to assume, at the instant of their formation, *a movement of rotation upon themselves,*—a movement which constantly takes place *in the same direction as that of the ring.* Moreover, as the ring, at the instant of its rupture, had still a remainder of velocity, the spheres to which it has given birth tend to fly off at a tangent; but, as on the other hand, the disc, turning in the alcoholic fluid, has impressed on this a movement of rotation, the spheres are especially carried along by this last movement, and revolve for some time round the disc. Those which revolve at the same time upon themselves, consequently then present the curious spectacle of *planets revolving at the same time upon themselves and in their orbits.* Finally—besides three or four large spheres into which the ring revolves itself, there are almost always produced one or two very small ones, which may thus be compared to satellites. The experiment which we have thus described, presents, as we see, an image in miniature of the formation of the planets, according to the hypothesis of Laplace, by the rupture of the cosmical rings attributable to the condensation of the solar atmosphere."* It must of course be admitted that the process of the experiment was of a reverse kind, and attended, as far as M. Plateau's description informs us, by slightly various effects; but the general reflection which it gives of the nebular cosmogony is certainly such as, with the other evidence, ought to go far to convince the unprejudiced of that cosmogony being true.

Surprising, then, as it may seem to the ordinary mind, it appears that we have approached to a distinct conception of the manner in which the

* See "Professor Plateau on the Phenomena presented by a free Liquid Mass withdrawn from the action of Gravity."—*Taylor's Scientific Memoirs*, November, 1844.

mighty spheres which voyage through fineless space have taken their present forms and arrangements. Proved the nebular cosmogony certainly is not, and yet it is brought perhaps as near to demonstration as any such fact can be, or as we have any reason to expect, considering the nature of the case. Some further support I trust to bring to it; but in the meantime, assuming its truth, let us see what idea it gives of the constitution of what we term the universe, of the development of its various parts, and of its original condition.

Reverting to a former illustration—if we could suppose a number of persons of various ages presented to the inspection of an intelligent being newly introduced into the world, we cannot doubt that he would soon become convinced that men had once been boys, that boys had once been infants, and, finally, that all had been brought into the world in exactly the same circumstances. Precisely thus, seeing in our astral system many thousands of worlds in all stages of formation, from the most rudimental to that immediately preceding the present condition of those we deem perfect, it is unavoidable to conclude that all the perfect have gone through the various stages which we see in the rudimental. This leads us at once to the conclusion that the whole of our firmament was at one time a diffused mass of nebulous matter, extending through the space which it still occupies. So also, of course, must have been the other astral systems. Indeed, we must presume the whole to have been originally in one connected mass, the astral systems being only the first division into parts, and solar systems the second.

According, then, to this hypothesis, the formation of worlds is a process involving time. It has long been and still is in progress, though, as far as respects the parts of space within our ken, much nearer to its completion than its commencement. Our own solar system is to be regarded as completed, supposing its perfection to consist in the formation of a series of planets, for there are mathematical reasons for concluding that Mercury is the nearest planet to the sun, which can, according to the laws of the system, exist. We have no means of judging of the seniority of systems; but it is reasonable to suppose that among the many that are complete, some are older than ours. There is, indeed, one piece of evidence for the probability of the comparative youth of our system, altogether apart from human traditions and the geognostic appearances of the surface of our planet. This consists in a thin nebulous matter, just alluded to under the name of the Zodiacal Light. It is diffused around the sun to nearly the orbit of Mercury, in a very oblately spheroidal form. Appearing to our naked eyes, at sunset, in the form of a cone projecting upwards in the line of the sun's path, it has been thought a residuum or last remnant of the concentrating matter of our system, thus indicating the comparative recentness of the principal events of our cosmogony. Supposing the surmise and inference to be correct, and they may be held as so far supported by more familiar evidence, we might with the more confidence speak of our system as not amongst the elder born of Heaven, but one whose various phenomena, physical and moral, as yet lay undeveloped, while myriads of others were fully fashioned, and in complete arrangement. Thus, in the sublime chronology to which we are directing our inquiries, we first find ourselves called upon to consider the globe which we inhabit as a child of the sun, older than Venus and her younger brother Mercury, but posterior in date of birth to Mars, Jupiter, Saturn, and Uranus; next to regard our whole system as probably of recent formation in comparison with many of the stars of our firmament. We must, however, be on our guard against supposing the earth as a recent globe in our ordinary conceptions of time. From evidence afterwards to be adduced, it will be seen that it cannot be less than many hundreds of centuries old. How much older Uranus may be, no one can tell, far less how much more aged may be many of the stars of our firmament, or the stars of other firmaments than ours.

The most striking induction pointed to by the nebular hypothesis, is, that the physical creation was in all its stages conducted on the principle of order or law. The nebulous matter collects round nuclei by virtue of the law of attraction. The agglomeration brings into operation another physical law, by force of which the separate masses of matter are either caused to rotate singly, or, in addition to that single motion, are set into a coupled revolution in ellipses. Next centrifugal force comes into play in connexion with the attractive law, and produces a progeny of planets all in the most complete relation as to their distances and times of revolution. The form which the single revolving globe takes is also precisely in obedience to a law. *We might, then, entirely dismiss the nebular theory, and still in the relations of the planets, and in the calculations as to their oblate spheroidality, we should have overpowering proof that the cosmical arrangements were produced in the way of natural law.*

It becomes of course important to settle the character of this so-called law. The word is, in reality, a term as much metaphorical as descriptive —one of convenience—designed only to express the invariable order observed in certain series of occurrences. We see the order and the invariableness; we conceive these to be impressed by a power external to the occurrence themselves. The occurrences, therefore, appear to be under regulations resembling those which men establish in their systems of social polity: they are, we say, under natural law. Such arrangements have long been ascertained to exist throughout the whole of the physical world; it has been the task of natural philosophy, by the aid of mathematics, and of chemistry, itself aided latterly by mathematics, to give us definite notions respecting these material laws. Thus we arrive at the idea of certain results, following from all the possible conditions and combinations, in which we can suppose matter to exist, and this at all times alike. We also see such results take place without any regard to magnitude of scale. The tear that falls to-day from childhood's cheek is globular, through a similar efficacy to that which made the sun and planets round. The rapidity of Mercury, as has been remarked, is greater than that of Saturn, for the same reason that when we wheel a ball round by a string, and cause the string to wind up round one of our fingers, the ball always flies quicker and quicker as the string is shortened. Two eddies in a stream fall into a mutual revolution at the distance of a few inches, through the same cause which makes a pair of suns link in mutual revolution at the distance of millions of miles. There is a sublime simplicity in this indifference of the grand regulations to the vastness or minuteness of the field of their operation,—as well as in the ceaseless perseverance of that operation from the earliest point to which we can extend our retrospect. Nor should it escape careful notice that the regulations on which all the laws of matter proceed, are established on a rigidly accurate mathematical basis. Proportions of numbers and geometrical figures rest at the bottom of the whole. Such considerations tend to raise our ideas with respect to the character of physical laws, which, however, we are not necessarily led to consider as the self-sufficient cause of all the phenomena of the universe. On the contrary, when we contemplate these beautiful arrangements, so admirably suited for the various ends which they serve, we unavoidably advance to the idea of an intelligence which has conceived and originated them, and by which they are sustained in action. We are, in short, brought at once to the acknowledgement of a Being, the Creator and Ruler of the universe, for whose *modes of action*, nature and natural law are but representative terms. That great Being, who shall say where his dwelling-place, or what his history! Man pauses breathless at the contemplation of a subject so much above his finite faculties, and only can wonder and adore!

CONSTITUENT MATERIALS OF THE EARTH AND OF THE OTHER BODIES OF SPACE.

The nebular hypothesis necessarily supposes matter to have originally formed one mass. We have seen that the same physical laws preside over the whole. Are we also to presume that the constitution of the whole was uniform?—that is to say, that the whole consisted of the same elements. It seems difficult to avoid coming to this conclusion, at least under the qualification that, possibly, various bodies, under peculiar circumstances attending their formation, may contain elements which are wanting, and lack some which are present, in others, or that some may entirely consist of elements in which others are entirely deficient.

What are elements? This is a term applied by the chemist to a limited number of substances, (fifty-four or fifty-five are ascertained,) which, in their combinations, form all the matters present in and about our globe. They are called elements, or simple substances, because it has hitherto been found impossible to reduce them into others, wherefore they are presumed to be the primary bases of all matters. It has, indeed, been surmised that these so-called elements are only modifications of a primordial form of matter, brought about under certain conditions; but if this should prove to be the case, it would little affect the view which we are taking of cosmical arrangements. Analogy would lead us to conclude that the modifications of the primordial matter, forming our so-called elements, are as universal, or as liable to take place everywhere, as are the laws of gravitation and centrifugal force. We must therefore presume that the gases, the metals, the earths, and other simple substances, (besides whatever more of which we have no acquaintance,) exist or are liable to come into existence under proper conditions, as well in the astral system which is thirty-five thousand times more distant than Sirius, as within the bounds of our own solar system or our own globe.

Matter, whether it consists of about fifty-five

ingredients, or only one, is liable to infinite varieties of condition under different influences. As a familiar illustration, water, when subjected to a temperature under 32° Fahrenheit, becomes ice; raise the temperature to 212°, and it becomes steam, occupying a vast deal more space than it formerly did. The gases, when subjected to pressure, become liquids; for example, carbonic acid gas, when subjected to a weight equal to a column of water 1230 feet high, at a temperature of 32°, takes this form: the other gases require various amounts of pressure for this transformation, but all appear to be liable to it when the pressure proper in each case is administered. Heat is a power greatly concerned in regulating the volume and other conditions of matter. The chemist will probably yet tell us what additional amount of heat would be required to vaporize all the water of our globe; how much more to disengage the oxygen which is diffused in nearly a proportion of one-half throughout its solids; and, finally, how much more would be required to cause the whole to become vaporiform, which we may consider as equivalent to its being restored to its original nebulous state. He may calculate with equal certainty, what would be the effect of a considerable diminution of the earth's temperature — what changes would take place in each of its component substances, and how much the whole would shrink in bulk.

The earth and all its various substances have at present a certain volume in consequence of the temperature which actually exists. If, then, we find that its matter and that of the associate planets was at one time diffused throughout the whole space now circumscribed by the orbit of Uranus, we cannot doubt, after what we know of the power of heat, that the nebulous form of matter was attended by the condition of a very high temperature. The nebulous matter of space, previously to the formation of stellar and planetary bodies, must have been a universal Fire Mist, an idea which we can scarcely comprehend, though the reasons for arriving at it seem irresistible. The formation of systems out of this matter implies a change of some kind with regard to the condition of the heat. Had this power continued to act with its full original repulsive energy, the process of agglomeration by attraction could not have gone on. We do not know enough of the laws of heat to enable us to surmise how the necessary change in this respect was brought about;

but we can trace some of the steps and consequences of the process. Uranus would be formed at the time when the heat of our system's matter was at the greatest, Saturn at the next, and so on. Now this tallies with the exceeding diffuseness of the matter of those elder planets, Saturn being not more dense or heavy than the substance cork. It may be that a sufficiency of heat still remains in those planets to make up for their distance from the sun, and the consequent smallness of the heat which they derive from his rays. And it may equally be, since Mercury is nearly thrice the density of the earth, that its matter exists under a degree of cold for which that planet's large enjoyment of the sun's rays is no more than a compensation. Thus there may be upon the whole a nearly equal experience of heat amongst all these children of the sun. Where, meanwhile, is the heat once diffused through the system, over and above what remains in the planets? May we not rationally presume it to have gone to constitute that luminous envelope of the sun, in which his warmth-giving power is now held to reside? It may have simply been reserved to constitute, at the last, a means of sustaining the many operations of which the planets were destined to be the theatre.

The tendency of the preceding considerations is to impress the notion that our globe is a specimen of all the similarly-placed bodies of space, as respects its constituent matter and the physical and chemical laws governing it, with only this qualification, that there are *possibly* shades of variation with respect to the component materials, and *undoubtedly* with respect to the conditions under which the laws operate, and consequently the effects which they produce. Thus, there may be substances here which are not in some of the other bodies, and substances here solid may be elsewhere liquid or vaporiform. We are the more entitled to draw such conclusions, seeing that there is nothing at all singular or special in the astronomical situation of the earth. It takes its place third in a series of planets, which series is only one of numberless other systems forming one group. It is strikingly—if I may use such an expression—a member of a democracy. Hence, we cannot suppose that there is any peculiarity about it which does not attach to multitudes of other bodies—in fact, to all that are analogous to it in respect to cosmical arrangements.

It therefore becomes a point of great interest—what are the materials of this specimen? What is the constitutional character of this object, which may be said to be a sample, presented to our immediate observation, of those crowds of worlds which seem to us as the particles of the desert sand-cloud in number, and to whose diffusion there are no conceivable local limits?

The solids, liquids, and aeriform fluids of our globe are all, as has been stated, reducible into fifty-five substances hitherto called elementary. Of these, forty are well-characterized metals, twelve non-metallic bodies, and the remaining three solid substances of intermediate character, which form a connecting link between the two great groups. Among the non-metallic elements, four—viz., oxygen, hydrogen, nitrogen, and chlorine, are permanently gaseous; bromine is fluid at common temperatures; and the remainder (with the exception of fluorine, which has never been isolated, and whose physical characters are consequently unknown) are solid.

The body oxygen is considered as by far the most abundant substance in our globe. It constitutes a fifth part of our atmosphere, eight-ninths of the weight of water, and a large proportion of every kind of rock in the crust of the earth. Hydrogen, which forms the remaining part of water, and enters into some mineral substance, is perhaps next. Nitrogen, of which the atmosphere is four-fifths composed, must be considered as an abundant substance. The metal silicium, which unites with oxygen in nearly equal parts to form silica, the basis of about a half of the rocks in the earth's crust, is, of course, an important ingredient. Aluminium, the metallic basis of alumina, a material which enters largely into many rocks, is another abundant elementary substance. So, also, is carbon, a small ingredient in the atmosphere, but the chief constituent of animal and vegetable substances, and of all fossils which ever were in the latter condition, amongst which coal takes a conspicuous place. The familiarly-known metals, as iron, tin, lead, silver, gold, are elements of comparatively small magnitude in that exterior part of the earth's body which we are able to investigate.

It is remarkable of the elementary substances that they generally exist in combination. Thus, oxygen and nitrogen, though in mixture they form the aerial envelope of the globe, are never found separate in nature. Carbon is pure only in the diamond. And the metallic bases of the earths, though the chemist can disengage them, may well be supposed unlikely to remain long uncombined, seeing that contact with moisture makes them burn. Combination and re-combination are principles largely pervading nature. There are few rocks, for example, that are not composed of at least two varieties of matter, each of which is again a compound of elementary substances. What is still more wonderful, with respect to this principle of combination, all the elementary substances observe certain mathematical proportions in their unions. When in the gaseous state, one volume of them unites with one, two, three, or more volumes of another, any extra quantity being sure to be left over, if such there should be. Combinations by weight are also governed by fixed and unchanging laws, of the greatest beauty and simplicity. It has hence been surmised that matter is composed of infinitely minute particles or atoms, each of which belonging to any one substance can only associate with a certain number of the atoms of any other. There are also strange predilections amongst substances for each other's company. One will remain combined in solution with another, till a third is added, when it will abandon the former and attach itself to the latter. A fourth being added, the third will perhaps leave the first, and join the new comer.

Such is an outline of the information which chemistry gives us regarding the constituent materials of our globe. How infinitely is the knowledge increased in interest, when we consider the probability of such being the materials of the whole of the bodies of space, and the laws under which these everywhere combine, subject only to local and accidental variations!

In considering the cosmogenic arrangements of our globe, our attention is called in a special degree to the moon.

In the nebular hypothesis, satellites are considered as masses thrown off from their primaries, exactly as the primaries had previously been from the sun. The orbit of any satellite is also to be regarded as marking the bounds of the mass of the primary at the time when that satellite was thrown off; its speed likewise denotes the rapidity of the rotatory motion of the primary at that particular juncture. For example, the outermost of the four satellites of Jupiter revolves round his body at the distance of 1,180,582 miles, showing that the planet was once about 3,675,501 miles in circumference, instead of being, as now, only

89,170 miles in diameter. This large mass took rather more than sixteen days six hours and a half (the present revolutionary period of the outermost satellite) to rotate on its axis. The innermost satellite must have been formed when the planet was reduced to a circumference of 309,075 miles, and rotated in about forty-two hours and a half.

From similar inferences, we find that the mass of the earth, at a certain point of time after it was thrown off from the sun, was no less than 482,000 miles in diameter, being sixty times what it has since shrunk to. At that time, the mass must have taken rather more than twenty-nine and a half days to rotate, (being the revolutionary period of the moon,) instead of, as now, rather less than twenty-four hours.

The time intervening between the formation of the moon and the earth's diminution to its present size, was probably one of those vast sums in which astronomy deals so largely, but which the mind altogether fails to grasp.

The observation made upon the surface of the moon by telescopes tends strongly to support the hypothesis as to all the bodies of space being composed of similar matters subject to certain variations. It does not appear that our satellite is provided with that gaseous envelope which, on earth, performs so many important functions. Neither is there any appearance of water upon the surface; yet that surface is, like that of our globe, marked by inequalities and the appearance of volcanic operations. These inequalities and volcanic operations are upon a scale far greater than any which now exist upon the earth's surface. Although, from the greater force of gravitation upon its exterior, the mountains, other circumstances being equal, might have been expected to be much smaller than ours, they are, in many instances, equal in height to nearly the highest of our Andes. They are generally of extreme steepness, and sharp of outline, peculiarities which might be looked for in a planet deficient in water and atmosphere, seeing that these are the agents which wear down ruggedness on the surface of our earth. The volcanic operations are on a stupendous scale. They are the cause of the bright spots of the moon, while the want of them is what distinguishes the duller portions, usually but erroneously called *seas.* In some parts, bright volcanic matter, besides covering one large patch, radiates out in long streams, which appear studded with subordinate *foci* of the same kind of energy. A large portion of the surface is covered with circular eminences, called Ring Mountains, of various diameters, from a quarter of a mile to several hundred miles, and in some places as close together as the circles on the surface of a boiling pot, which they in no small degree resemble. Some even intrude upon and obliterate portions of the neighbouring circles, thus leading to the idea of *date*, or a succession of events on the moon's surface. Generally, in the centre, there is a mount, which appears to be connected, in the way of cause, with the annular eminence, beyond which again vast boulder-like masses are in some instances seen scattered. What, however, most strikes the senses of an observer, is the vast profundity of some of the pits between the ring and the inner mount; in one case, this is reckoned to be not less than 22,000 feet, or twice the height of Ætna.

These characteristics of the moon forbid the idea that it can be at present a theatre of life like the earth, and almost seem to declare that it never can become so. But it is far from unlikely that the elements which seem wanting may be only in combinations different from those which exist here, and may yet be developed as we here find them. Seas may yet fill the profound hollows of the surface; an atmosphere may spread over the whole. Should these events take place, meteorological phenomena, and all the phenomena of organic life, will commence, and the moon, like the earth, will become a green and inhabited world.*

It is unavoidably held as a strong proof in favour of any hypothesis, when all the relative phenomena are in harmony with it. This is eminently the case with the nebular hypothesis, for

* Among the most extraordinary phenomena of natural science must be placed those relating to the fall of *meteoric stones.* The fact itself, so long doubted, has now been established by an accumulation of the most positive and unexceptionable evidence. The stones have been seen to fall, and taken up in a still heated state;—there can be no manner of doubt about the fact, although the explanation is extremely difficult. All these stones are found on examination to resemble each other in their general characters; they usually consist of an earthy material, having disseminated through its substance globules and small masses of metallic iron containing nickel in the state of alloy. The stones are often covered by a thin vitreous crust, as if partial fusion had commenced. It is well known, also, that large masses of soft, malleable iron, also containing nickel, are found in several places far removed from each other, lying loose upon the earth, as in South America and in Siberia, and no doubt can exist of the meteoric origin of these masses. It has been conjectured that these meteoric stones proceed from the

here the associated facts cannot be explained on any other supposition. We have seen reason to conclude that the primary condition of matter was that of a diffused mass, in which the component molecules were probably kept apart through the efficacy of heat; that portions of this agglomerated into suns, which threw off planets; that these planets were at first very much diffused, but gradually contracted by cooling to their present dimensions. Now, as to our own globe, there is a remarkably distinct memorial of the original high temperature of the materials, in the store of

moon, having been shot out from volcanoes with such violence as to be brought within the reach of the earth's attraction. A view now more generally received supposes the existence in space of very numerous small bodies, moving in more or less regular orbits around the sun and larger planets, which at certain periods undergo such perturbation that their motion becomes completely deranged, and they at length fall upon the surface of the earth or other planet, whose attraction has been the exciting cause of the derangement of their orbits. Whatever may be their real origin, they are by common consent looked upon as foreign to the earth: their physical constitution is completely different from any known minerals. But what is exceedingly remarkable, and particularly worthy of notice as strengthening the argument that all the members of the solar system, and perhaps of other systems, have a similar constitution, *no new elements* are found in these bodies; they contain the ordinary materials of the earth, but associated in a manner altogether new, and unlike anything known in terrestrial mineralogy.—*Note by a Correspondent.*

heat which still exists in the interior. The immediate surface of the earth, be it observed, exhibits only the temperature which might be expected to be imparted to such materials by the heat of the sun. There is a point a very short way down, but varying in different climes, where all effect from the sun's rays ceases. Then commences a temperature from an entirely different cause, one which evidently has its source in the interior of the earth, and which regularly increases as we descend to greater and greater depths, the rate of increment being, in some places, about one degree Fahrenheit for every hundred feet; and of this high temperature there are other evidences, in the phenomena of volcanoes and thermal springs, as well as in what is ascertained with regard to the density of the entire mass of the earth. This approximates five and a half times the weight of water; but the actual weight of the principal solid substances composing the outer crust is as two and a half times the weight of water; and this, we know, if the globe were solid and cold, should increase vastly towards the centre, water acquiring the density of quicksilver at 362 miles below the surface, and other things in proportion, and these densities becoming much greater at greater depths;

so that the entire mass of a cool globe should be of a gravity infinitely exceeding five and a half times the weight of water. The only alternative supposition is, that the central materials are greatly expanded or diffused by some means; and by what means could they be so expanded but by heat? Indeed, the existence of this central heat, a residuum of that which kept all matter in a vaporiform chaos at first, is amongst the most solid discoveries of modern science,* and the support which it gives to the nebular hypothesis, is highly important. We shall hereafter see what have been supposed by some to be traces of an operation of this heat upon the surface of the earth in very remote times; an effect, however, which has long passed entirely away.

* The researches on this subject were conducted chiefly by the late Baron Fourier, perpetual secretary to the Academy of Sciences of Paris. See his *Théorie Analytique de la Chaleur*, 1822.

THE EARTH FORMED—ERA OF THE PRIMARY ROCKS.

ALTHOUGH the earth has not been actually penetrated to a greater depth than three thousand feet, the nature of its material can, in many instances, be inferred for the depth of many miles by other means of observation. We see a mountain composed of a particular substance, with strata, or beds of other rock, lying against its sloped sides; we, of course, infer that the substance of the mountain dips away under the strata which we see lying against it. Suppose that we walk away from the mountain across the turned up edges of the stratified rocks, and that for many miles we continue to pass over other stratified rocks, all disposed in the same way, till we at length come to a place where we begin to cross the opposite edges of the same beds. We then pass over these rocks, all in reverse order, till we come to another extensive mountain composed of similar material to the first, and shelving away under the strata in the same way. We should then infer that the stratified rocks occupied a basin formed by the material of these two mountains, and by calculating the thickness right through these strata, could say to what depth the rock of the mountain extended below. By such means, the kind of rock existing many miles below the surface can often be inferred with considerable confidence.

The interior of the globe has now been inspected in this way in many places, and a tolerably distinct notion of its general arrangements has consequently been arrived at. It appears that the basis rock of the earth, as it may be called, is of hard texture, and crystalline in its constitution. Of this rock, granite may be said to be the type, though it runs into many varieties. Over this, except in the comparatively few places where it projects above the general level in mountains, other rocks are disposed in sheets or strata, with the appearance of having been deposited originally from water. But these last rocks have nowhere been allowed to rest in their original arrangement.

Uneasy movements from below have broken them up in great inclined masses, while in many cases there has been projected through the rents rocky matter more or less resembling the great inferior crystalline mass. This rocky matter must have been in a state of fusion from heat at the time of its projection, for it is often found to have run into and filled up lateral chinks in these rents. There are even instances where it has been rent again, and a newer melted matter of the same character sent through the opening. Finally, in the crust as thus arranged, there are, in many places, chinks, and what are usually called veins, containing metal. Thus, there is first a great inferior mass, composed of crystalline rock, and probably resting immediately on the fused and expanded matter of the interior: next, layers or strata of aqueous origin; next, irregular masses of melted inferior rock that have been sent up volcanically and confusedly at various times amongst the aqueous rocks, breaking up these into masses, and tossing them out of their original levels. This is an outline of the arrangements of the crust of the earth, as far as we can observe it. It is, at first sight, a most confused scene; but after some careful observation, we detect in it a regularity and order from which much instruction in the history of our globe is to be derived.

The deposition of the aqueous rocks, and the projection of the volcanic, have unquestionably taken place since the settlement of the earth in its present form. They are indeed of an order of events which we see going on, under the agency of intelligible causes, down to the present day. We may therefore consider them generally as comparatively recent transactions. Abstracting them from the investigations before us, we arrive at the idea of the earth in its first condition as a globe of its present size—namely, as a mass, externally at least, consisting of the crystalline kind of rock, with the waters of the present seas and the present atmosphere around it, though these were perhaps in different conditions, both as to temperature and their constituent materials, from what they now are. We are thus to presume that that crystalline texture of rock which we see exemplified in granite is the condition into which the great bulk of the solids of our earth were agglomerated directly from the nebulous or vaporiform state. It is a condition eminently of combination, for such rock is invariably composed of two or more of four substances—filspar, mica, quartz, and hornblende—which associate in it in distinct crystals, and which are themselves each composed of a group of the simple or elementary substances.

Judging from the results and from still remaining conditions, we must suppose that the heat retained in the interior of the globe was more intense, or had greater freedom to act in some places than in others. These became the scenes of volcanic operations, and in time marked their situations by the extrusion of traps and basalts from below—namely, rocks composed of the crystalline matter fused by intense heat, and developed in various conditions, according to particular circumstances; some, for example, reaching the surface either under the water or atmosphere, and others not, which contingencies are found to have made considerable difference in their texture and appearance. The great stores of subterranean heat also served an important purpose in the formation of the aqueous rocks. These rocks might, according to Sir John Herschel, become subject to heat in the following manner: —While the surface of a particular mass of rock forms the bed of the sea, the heat is kept at a certain distance from that surface by the contact of the water; philosophically speaking, the mass radiates away the heat into the sea; to resort to common language, it is cooled a good way down. But when new sediment settles at the bottom of that sea, the heat rises up to what was formerly the surface; and when a second quantity of sediment is laid down, it continues to rise through the first of the deposits, which then becomes subjected to those changes which heat is calculated to produce. This process is precisely the same as that of putting additional coats upon our own bodies; when, of course, the internal heat rises through each coat in succession, and the third (supposing there is a fourth above it) becomes as warm as perhaps the first originally was.

In speaking of sedimentary rocks, we may be said to be anticipating. It is necessary, first, to show how such rocks were formed, or how stratification commenced.

Geology tells us as plainly as possible, that the original crystalline mass was not a perfectly smooth ball, with air and water playing round it. There were irregularities in the surface,—irregularities, trifling, perhaps, compared with the whole bulk of the globe, but probably larger than any which now exist upon it. These irregularities might be occasioned by inequalities in the cooling of the substance, or by accidental and local sluggishness of the materials, or by local effects of the concentrated internal heat. From whatever cause they arose, there they were—granitic mountains, interspersed with seas which sunk to a great depth, and by which, perhaps, the mountains were wholly or partially covered. Now, it is a fact of which the very first principles of geology assure us, that the solids of the globe cannot for a moment be exposed to water, or to the atmosphere, without becoming liable to change. They instantly begin to be worn down. This operation, we may be assured, proceeded with as much certainty in the earliest ages of our earth's history, as it does now, but probably upon a much more magnificent scale. The matters worn off, being carried into the neighbouring depths, and there deposited, became the components of the earliest stratified rocks, the first series of which is the *Gneiss and Mica Slate System,* or series, an example of which is exposed to view in the Highlands of Scotland. We have evidence that the earliest strata were formed in the presence of a stronger degree of heat than what operated in subsequent stages of the world, for the laminæ of the gneiss and of the mica and chlorite schists are contorted in a way which could only be the result of a very high temperature. It appears as if the seas in which these deposits were formed, had been in the troubled state of a caldron of water nearly at boiling heat. Such a condition would undoubtedly add greatly to the disintegrating power of the ocean.

The earliest stratified rocks contain no minerals which are not to be found in the primitive granite. They are the same in material, but only changed into new forms and combinations. But how comes it that some of them are composed almost exclusively of one of the materials of granite; the mica schists, for example, of mica—the quartz rocks, of quartz, &c.? For this there are both chemical and mechanical causes. Suppose that a river has a certain quantity of material to carry down, it is evident that it will soonest drop the larger particles, and carry the lightest furthest on. To such a cause is it owing that some of the materials of the worn-down granite have settled in one place and some in another.* Again, some of these materials must be presumed to have been

* De la Beche's Geological Researches.

in a state of chemical solution in the primeval seas. It would be of course in conformity with chemical laws, that certain of these materials should be precipitated singly, or in modified combinations, to the bottom, so as to form rocks by themselves.

COMMENCEMENT OF ORGANIC LIFE—SEA PLANTS, CORALS, ETC.

THE group of rocks placed above the Primary, bears the general name of *Silurian*, from their being presented on the surface in a portion of Western England formerly occupied by a people whom the Roman historians call Silures. It is also developed very extensively in Northern Europe, and in North America. The group is divided by geologists into two distinct formations, the Lower and Upper.

Hitherto nothing has been said of the fossils which constitute so important a part of geological science. It is now to be observed that, from an early portion of the rock series to its close, the mineral masses are found to enclose remains of the organic beings (plants and animals) which flourished upon earth during the time when those were forming; and these organisms, or such parts of them as were of sufficient solidity, have been in many instances preserved with the utmost fidelity, although for the most part converted into the substance of the enclosing mineral. The rocks may be said thus to form an organic history of our globe, from perhaps near the commencement of life upon its surface to the present time. This is a piece of knowledge entirely new to man, and it may be safely said that he has never made a merely intellectual acquisition of a more interesting or remarkable nature. I am to trace this history as well as existing materials will permit.

The first leaves of the Stone Book, like those of many written histories, tell a somewhat obscure tale. It is impossible to say precisely when life began upon our planet, or of what forms it consisted. It only appears certain that, in these early ages of the earth's history, the plants and animals which existed were, generally speaking, of humble kinds, and exclusively marine. Here it may be necessary to remark that a plant or animal is said to be humble, when its organization is of a simple kind, subservient to a comparatively narrow range of functions, and suited to a comparatively limited field of existence. Confervæ, fuci, algæ, mosses, are of this grade in the vegetable kingdom. In the department of zoology, the lowest grade is composed of what Cuvier calls *Radiata*, as infusory animalcules, sponges, corals, and star-fishes. Above these, and on one level, are *Mollusca*, including shell-fish, cuttle-fishes, etc., and *Articulata*, to which belong crustaceous animals, insects, and spiders. A grand and crowning division are *Vertebrata*, or backboned animals, composed of four classes — fish, reptiles, birds, and mammalia. Now, one fact is certain, that the sea of the Lower Silurian era, a vast space of time, contained no fish, nor any other vertebrate animals. This is found to be the case in England, Russia, and America alike. The animals actually presented are confined to the three lowest divisions of the animal kingdom, and generally are of the humblest forms of those divisions. The plants are also restricted to a very lowly grade in their department. It is here proper to remark that the kind of animals inhabiting particular parts of the present seas is determined by peculiar circumstances with respect to depth of water, currents, and temperature. There is also a variation of species in different parts of the earth, even in circumstances otherwise perfectly parallel. We are therefore to expect that there should be some differences among the fossils of countries widely divided, and even in different districts of one region. From this explanation, the reader will know what importance to attach to the uniformity and what to the discrepancy of fossils in the Lower Silurians of different countries.

In those of Western England, the lowest bands, besides inferior animals, give specimens of *orthis*, a family belonging to a destructive class of molluscs, (the cephalopoda,) which are ranked as the highest of that sub-kingdom. With these are inferior molluscs and radiated animals, besides crustaceans. In America, again, the lowest fossiliferous rocks as yet detected present no molluscs of so high a grade. Indeed, here, as in other regions, it is admitted that the predominating animals are shell-fish of the class brachiopoda, and crustaceans of the trilobite family. In Russia, nearly one half of the species are identical with those in England, and we have the opinion of Mr. Lyell that the proportion in America identical with other Silurian regions is not less than we might anticipate "from the laws governing the distribution of living invertebrate animals."

To descend to a few particulars respecting this early fauna:—Of the Radiates, the *polypiaria* include various forms of those extraordinary animals (corallines) which abound to such a degree in tropical seas of the present day, often obstructing the course of the mariner, and even laying the foundations of new continents. The *crinoids* are an early and simple form of the large family of echinodermata (star-fishes), also very abundant in the present day; the animal, though composed of innumerable minute calcareous masses, connected by a gelatinous substance, is merely a stomach surrounded by tentacula, to provide itself with food, and mounted upon a many-jointed stalk, so as to bear a considerable resemblance to a flower growing on its stem. There is also in the lower fossiliferous rocks a vast multitude of zoophytes allied to the sea-pens of modern seas, a family of animals usually inhabiting mud and slimy sediment in deep water. Of the crustacea of the system, the most remarkable are *trilobites*—animals which continued to flourish in a great variety of species throughout several of the subsequent rock-formations, but which are now only faintly represented in a few obscure species. Some curious inferences have been made by Dr. Buckland, from the prominent facet-covered eyes with which this creature was furnished, indicating that the sea in which it lived was a clear medium, as existing seas generally are, and that light was the same in character in those inconceivably remote ages as it is now.

We are called upon, however, to believe that the few species of radiates, shell-fish, and trilobites, which have left their remains in this group of strata, were not the first sole examples of life which existed upon the earth. We see such animals in the present day requiring smaller and simpler animals for their subsistence; and tracing, again, the economy of these smaller animals, we find that they depend for nutriment either upon animals still more minute than themselves, or upon some equally small and impalpable forms of vegetation, which the bountiful hand of Nature has placed within their reach. The crustacea, then, the mollusks, and radiates of the Lower Silurian era, necessarily imply the previous, or at least contemporary existence of certain humbler forms of life, vegetable as well as animal; forming a scene, indeed, much like what is found in seas of the present day, excepting that neither fishes nor any higher vertebrata as yet roamed through the marine wilds. That no very distinct or abundant remains of such plants and animals have been preserved for our examination, is precisely what might have been expected, for their forms were of such a soft consistence, as not to have had more than a slight chance of leaving memorials of themselves in rocks of any kind, more particularly rocks which are believed to have been subjected, in their formation, to an unusual degree of heat. It is not conceivable that confervæ, or any other simple forms of marine vegetation, that soft animalcules, or spongiæ, or acalephæ, could have preserved shape or consistence in the mud at the bottoms of those seas, till that mud had been baked into rock. There are, nevertheless, some appearances in the primary rocks, which may be said to betray the existence of organic life in that age. In those rocks, "fragments apparently organic, and resembling the cases of infusoria, [shelled animalcules,]" have been detected.* This is also what might have been expected, seeing that these infusoria, though of one of the humblest forms of animal being, possessed hard parts capable of being preserved. The existence of remains of animals in the primary rocks has been inferred by

* Ansted's Geology, ii. 60.

Braconnot, from his finding ammoniacal and combustible products on distilling portions of them in porcelain retorts. It is also to be remarked, that amidst the primary rocks there are a few patches of limestone, forming rather an exceptive or local, than a general phenomenon. Limestone (carbonate of lime) contains an element (carbon) which plants take in from the atmosphere, where it is a subordinate ingredient; marine polypes also appropriate it, in connexion with lime, from the waters of the ocean, provided it be there in solution; and this compound substance do these animals deposit in the far extending masses which have been alluded to. It is fully ascertained of many strata of limestone higher in the series, that they are merely coralline accumulations, changed by subjection to heat and pressure. It is evident, then, that though we do not find incontestable relics of distinct animals or plants below the Lower Silurians, there is much reason to suppose that life began at an earlier period, and in forms of a kind humbler than many of those found in that portion of the rock series.*

* It has been stated that the Gneiss and Mica system in Bohemia includes some seams of grauwacké, in which are organic remains. To this announcement British geologists have not as yet attached much importance. Dr. M'Culloch found, as he

The necessity of abstaining from rash assumptions with regard to the first forms of life, is fully shown by the history of the early vegetable fossils. It was only a few years ago supposed that animals had preceded plants, because their remains are found in the Lower Silurians, where at that time no distinct traces of vegetation had been detected. Now the absurdity of such a supposition from merely negative evidence is clearly shown; for fucoids are announced from both Russia and America; in the former country, below the position of any animal remains,* and, in the latter, in the very first fossiliferous sub-group, the Potsdam Sandstone. In the Lower Silurians of Southern Sweden, there are not only distinct impressions of such plants, but Professor Forchhammer speaks of courses of true coal, composed, as he thinks, of sea-weed, and gives an opinion that the alum slate of that country owes its combustible character to the carbon, sulphur, and potash, derived from marine vegetation.†

believed, fossil orthocerata in the Gneiss tract of Loch Eribol, in Sutherland; but Messrs. Sedgwick and Murchison, on a subsequent search, could not verify the discovery.

* Mr. Murchison, at British Association, 1844; see Report in Literary Gazette.

† Report, Brit. Assoc., 1844.

It is to be remarked of the fossils of this early period, that though they can readily be referred to existing *orders*, the species and even genera to which they belonged are no longer found on earth; nay, almost the whole had become extinct before the next group of strata was formed. Such changes of species we shall find to be of frequent occurrence throughout the subsequent ages. It may also be observed that the fossils of the slate system are at once few in number of species, and rare as individual specimens. In England, there are not more than thirty of them altogether, and it has been said that a geologist may travel thirty miles before he would collect specimens to an equal amount.

Ascending to the next group of rocks (Upper Silurians) we find fossils much more abundant, and also more various; while some important additions are made. The general forms are similar to those seen in the previous era, but most of the specific characters—the peculiar characters which form with naturalists reasons for assigning a peculiar name, as of a distinct species—are changed; only a few of the Lower Silurian creatures survive (at least in their original forms) into this era. For these changes, it is believed that alterations in the field of existence—the sea-bottom—formed a cause. From the lowest beds upwards, there are polypiaria; trilobites; brachiopodous mollusks, a vast number of genera, (including terebratula, pentamerus, spirifer, orthis, leptæna;) gasteropoda; cephalopoda, of several orders and many genera, (including turritella, orthoceras, nautilus, bellerophon.) These last animals, of which we have still conspicuous specimens in the nautilus and cuttle-fish, were eminently carnivorous, and must have acted the part of a police in keeping down the redundant life of the early seas.

A little above the Llandeilo rocks of this formation, there have been discovered certain convoluted forms, which are now established as marine annelids or sea-worms, a tribe which still exists in great number and varieties, either swimming freely in the sea, or inhabiting the sand of its bottom, or clefts in rocks and other submarine places of a sheltered kind. The discovery of such animals at this portion of the rock series, is important, as the order to which they belong (*Annelides*), from their having a high circulatory system, with red blood, (though inferior in many respects to other articulated animals,) are regarded as connecting them with some of the humblest of the class of fishes.

The Wenlock limestone is the most remarkable amongst all the rocks of the Silurian system for organic remains. Many slabs of it are wholly composed of corals, shells, and trilobites, held together by an argillo-calcareous cement. It contains many genera of crinoidea and polypiaria, and there is little reason to doubt that some beds of it are wholly the production of the latter creatures, or are, in other words, coral reefs transformed by heat and pressure.

In this formation, also, we have the first examples which have been discovered of fish; but they are generally of obscure character. It may here be remarked that the caution as to negative evidence hardly applies in this case, as the Lower Silurians have been so carefully examined in various parts of the world, that any such fossils, if they had existed, at least of genera containing parts of any solidity, could not have failed to be detected. We have then, in this history, not only a time when there were no land plants or animals, which might have been the simple result of there being as yet no land, but an incontestable record that the seas were for ages devoid of fish, although we can see no reason for its being unable to support such tenants. Next to the generally humble character of the Silurian fauna, this is the most remarkable fact as yet presented to us by these curious chronicles of organic creation.

ERA OF THE OLD RED SANDSTONE—FISHES ABUNDANT.

We advance to a new chapter in this marvellous history—the era of the *Old Red Sandstone System.* This term is applied to a series of strata, of enormous thickness in the whole mass, largely developed in Herefordshire, Shropshire, Worcestershire, and South Wales; also in the counties of Fife, Forfar, Moray, Cromarty, and Caithness; and in Russia and North America, if not in many other parts of the world. The particular strata forming the system are somewhat different in different countries; but there is a general character to the extent of these being a mixture of flagstones, marly rocks, and sandstones, usually of a laminous structure, with conglomerates. In the conglomerates, of great extent and thickness, which form, in at least one district, the basis or leading feature of the system, inclosing water-worn fragments of quartz, and other rocks, we have evidence of the seas of that period having been subjected to a violent and long-continued agitation, probably from volcanic causes. The upper members of the series bear the appearance of having been deposited in comparatively tranquil seas. The English specimens of this system show a remarkable freedom from those disturbances which result in the interjection of trap; and they are thus defective in mineral ores. In some parts of England the old red sandstone system has been stated as 10,000 feet in thickness.

In this era, most of the forms of life which existed in the Silurian era are found no more; only about one hundred out of eight are continued. We have, however, the same orders of marine creatures, zoophyta, polypiaria, mollusca, crustacea; to these are added numerous fishes, some of which are of extraordinary and surprising forms. Several of the strata are crowded with remains of fish, showing that the seas in which those beds were deposited had swarmed with that class of inhabitants. The investigation of this system is recent; but already M. Agassiz has ascertained about twenty genera and thrice the number of species. And it is remarkable that the Silurian fishes are here only represented in genera; the whole of the *species* of that time had already been changed. Even throughout the sub-groups of the system itself, the species are changed; and these are phenomena observed throughout all the subsequent systems or geological eras; apparently arguing, that during the deposition of all the rocks, a gradual change of physical conditions was constantly going on. A varying temperature, or a varying depth of sea, would at present be attended with similar changes in marine life; and by analogy, we are entitled to assume that such variations in the ancient seas might be amongst the causes of that constant change of genera and species in the inhabitants of those seas to which the organic contents of the rocks bear witness.

The predominating fishes of this system, and the only ones which (as far as fossils show) existed for some ages, are arranged by M. Agassiz in two orders, with a regard to their external covering, which that naturalist holds to be, in fishes, a reflection of the internal organization. Both orders, it is to be remarked at the very first, are manifestly of an inferior character to the two other orders which afterwards came into existence, and still are the principal fishes of our seas, these being covered by true scales, and respectively named ctenoid and cycloid, from the forms of that part of their organization. The two orders of early fish are covered with integuments considerably different in character; the one (*placoids*) with irregular enamelled plates, the other (*ganoids*) with regular enamelled scales, the first being not placed over each other, as scales are, but laid edge to edge, in the manner of a pavement. These characters, according to M. Agassiz, were accompanied by a rudimentary or cartilaginous skeleton, while the ctenoids and cycloids possess an osseous structure.

The *cephalaspis* very much resembles in form the asaphus, a trilobite of lower formations; having a longish tail-like body inserted within the cusp of a large crescent-shaped head, somewhat like a saddler's cutting-knife. The body is covered with strong plates of bone, enamelled, and the head was protected on the upper side with one large plate, as with a buckler—hence the name, implying *buckler-head*. A range of small fins conveys the idea of its having been as weak in motion as it is strong in structure. In the *coccosteus*, the outline of the body is of the form of a short thick coffin, rounded, covered with strong bony plates, and terminating in a long tail, which seems to have been the sole organ of motion. While the tail establishes this creature among the vertebrata and the fishes, its teeth, chiselled, as it were, out of the solid bone of the jaw, like the nippers of a lobster, and its mouth opening vertically, contrary to the usual mode of the vertebrata, forcibly suggests an alliance, which however may be fanciful, to the crustaceans. The *pterichthys* has also strong bony plates over its body, arranged much like those of a tortoise, and has a long tail; but its most remarkable feature, and that which has suggested its name, is a pair of long and narrow wing-like appendages attached to the shoulders, which the creature is supposed to have erected for its defence when attacked by an enemy.

A group of ganoids seem to have been the police of their day, possessing a powerful development of sharp conical teeth situated on the margin of the jaws. One genus, the *holoptychius*, introduced near the close of the Old Red era, and passing up into the next, presents a flat oval form, measuring in one specimen thirty inches by twelve, with a covering of strong plates, wavily grooved and overlapping each other, the head forming only a slight rounded projection from the general figure. We here find the first examples of animals which may be called *large*. In the strata of this formation at Dörpat, there are gigantic bones, which were at first thought to belong to reptiles, but have since been ascertained to be remains of fishes, leading to the conjecture that the animals to which they appertained could not be less than thirty-six feet long.*

M. Agassiz has lately announced nine genera of sharks of the division *Cestraceon* in the Old Red sandstones of Russia, and one example of such a family is said to have been found in the shales alternating with the Wenlock limestone, a portion of the Upper Silurian formation. It is in this voracious family that we see the placoids represented in modern seas; the ganoids are all but unrepresented in our time. Of both classes, one invariable peculiarity has attracted much attention. "In all recent fish, with the exception of the shark family, the sturgeon, and the bony pike, the vertebral

* The head fountain of information on the early fishes is M. Agassiz's Fossil Ichthyology, a splendid but not readily accessible book. For more popular descriptions, reference may be made to "New Walks in an Old Field, by Hugh Miller," Edin., 1842, and to Jameson's Journal, July and October, 1844. See also the excellent manual of Professor Ansted.

column terminates at the point where the caudal fin is given off, and this fin is expanded above and below the body, forming what is called a *homocercal* tail. In all those, without exception, which have been found in strata of the Palæozoic period, [placoids and ganoids,] the caudal fin is *heterocercal*, being formed of two unequal branches, the upper one expanded immediately from the vertebral column, while the lower one is given off at a point some distance from the extremity."* Now it is a remarkable fact, that this one-sided tail is a peculiarity in the more perfect fishes (as the salmon) at a certain stage in their embryonic history; as is also the inferior position of the mouth, peculiar to the early fishes. More than this: in the earlier periods of embryonic life, there is no vertebral column. This organ is represented in embryos by a gelatinous cord, called the dorsal cord, which in maturity disappears as the vertebræ are formed upon it. M. Agassiz has satisfied himself that this was the nature of the organization of the early fishes, as it is that of the sturgeon of the present seas. It is not premature to remark how broadly these facts hint at a parity of law affecting the progress of general creation, and the progress

* Ansted's Geology, i. 185.

of an individual fœtus of one of the more perfect animals.* Another feature of the placoids, bringing them down towards the level of an inferior portion of the animal kingdom, is the distinct marks which the dermal plates bear, in many specimens, of processes for muscular attachments. This suggests a peculiarity of crustacean animals, and powerfully hints that the cartilaginous skeleton had not been, as in higher vertebrata, the grand support of the frame, and the basis of its strength.

It is remarked of the fossils of this era, as of the preceding, that they vary locally, as far as might be expected from what we see of the distribution of animal life in the present times; but, throughout the distant parts of the earth where Old Red strata are found, the general characters of animal and also vegetable life are nearly the same. It is further observed that whatever particular family is continued with little change through a succession of strata, is also amongst those most widely extended over the world. It is the opinion of M. Brongniart, who has distinguished himself by his investigation of vegetable fossils, that the fuci of these early seas indicate a higher temperature than now

* See Appendix, B.

prevails at many of the places where they are found. He regards this as a proof of the more equable diffusion of a tropical climate in ancient times, and attributes it to the action of the internal heat of the earth. The early animals are not so uniform over large geographical areas as the plants. M. Agassiz surmises, from an examination of the fishes of the ancient seas, that the ocean did not at first contain much salt, but gradually acquired its present infusion of that material; a theory, it may be remarked, which derives support from a recent suggestion, that the salt of the sea has been mainly brought thither, in the course of time, by rivers, washing it in particles out of the land, in common with other detritus, while it is obvious that rain does not restore it.* It is easy to suppose a comparative absence of salt in the early ocean affecting animal and vegetable marine life in different ways and degrees.

As yet—overlooking possible exceptions of a narrow and dubious kind—we meet with no traces of land plants: remains of terrestrial animals have not even been suspected. This exclusively marine character of the flora and fauna of the ante-Carboniferous ages is usually thought to betoken the

* See Fownes's Actonian Prize Essay.

non-existence of dry land. And for this view there is some apparent support, in the observations which have been made on the history of mountains. The fact that early strata, though they must have been formed in a horizontal position, are usually found tilted up along the slopes of the granitic masses which form the nuclei or axes of great mountain ranges, implies of course that these mountain masses have been protruded from below at a period subsequent to the deposition of the strata; and thus, it is thought, we see causes for an emergence of land at a time following the formation of the primary series of rocks. But, on the other hand, it is not easy to understand how the vast disintegration which produced the early stratified rocks should have been effected, if there were no emerged masses, as the wearing down of rocks chiefly takes place at the point where land and water meet, and in a very small degree within the bosom of the waves. It therefore seems necessary to presume that dry land existed in these early ages, though, from whatever cause, it had produced little or no vegetation, and sustained no animals. And in this case, the protrusion of the granitic nuclei must be held to be an indifferent matter. It is, however, worthy of notice that there is no part of geological science more clear than that which refers to the ages of mountains. It is as certain that the Grampian mountains of Scotland are older than the Alps and Apennines, as it is that civilization had visited Italy, and had enabled her to subdue the world, while Scotland was the residence of "roving barbarians." The Pyrenees, Carpathians, and other ranges of continental Europe, are all younger than the Grampians, or even the insignificant Mendip Hills of southern England. Stratification tells this tale as plainly as Livy tells the history of the Roman republic. It tells us—to use the words of Professor Phillips —that at the time when the Grampians sent streams and detritus to straits where now the valleys of the Forth and Clyde meet, the greater part of Europe was a wide ocean.

The last three systems are of great thickness, and of extensive distribution; not only implying a vastness and a uniformity of agency, but clearly demonstrating the lapse of long eras, during which repeated changes of relative level between the solid and watery surfaces had taken place, thus progressively affecting those conditions which regulate both the distribution and kind of organic existence.

SECONDARY ROCKS.

ERA OF THE CARBONIFEROUS FORMATION.

COMMENCEMENT OF LAND PLANTS.

THE Secondary Rocks, in which our further researches are to be prosecuted, consist of a great and varied series, resting, generally unconformably, against flanks of the upturned primary rocks, sometimes themselves considerably inclined, at others, forming extensive basin-like beds, nearly horizontal; in many places much broken up and shifted by disturbances from below. They have all been formed out of the materials of the older rocks, by virtue of the wearing power of air and water, which is still every day carrying down vast quantities of the elevated matter of the globe into the sea. But the separate strata are each much more distinct in the matter of its composition than might be expected. Some are siliceous or arenaceous, (sandstones,) composed mainly of fine grains from the quartz rocks — the most abundant of the primary strata. Others are argillaceous—clays, shales, &c., chiefly derived, probably, from the slate beds of the primary series. Others are calcareous, derived partly from the limestones, and partly eliminated from the waters of the ocean by organic life. As a general feature, they are softer and less crystalline than the primary rocks, as if they had endured less of both heat and pressure than the senior formation. There are beds (*coal*) formed solely of vegetable matter, and some others in which the conspicuous ingredient is a carbonate of iron, (*the black band.*) The secondary rocks are quite as communicative with regard to their portion of the earth's history as the primitive were.

The first, or lowest, group of the secondary rocks is called the *Carboniferous Formation*, from the remarkable feature of its numerous interspersed beds of coal. It commences with the beds of the *mountain limestone*, which, in some situations, as in Derbyshire and Ireland, are of great thickness, being alternated with chert, (a siliceous sandstone,) sandstones, shales, and beds of coal, generally of the harder and less bituminous kind, (*anthracite*,) the whole being covered in some places by the millstone grit, a siliceous conglomerate, composed of the detritus of the primary rocks. The mountain limestone, attaining in England to a depth of eight hundred yards, greatly exceeds in volume any of the primary limestone beds, and shows an enormous addition of power to the causes formerly suggested as having produced this substance. In fact, distinct remains of corals, crinoidea, and shells, are so abundant in it, as to compose three-fourths of the mass in some parts. Above the mountain limestone commence the more conspicuous *coal beds*, alternating with sandstones, shales, beds of limestone, and ironstone. Coal is altogether composed of the matter of a terrestrial vegetation, transmuted by putrefaction of a peculiar kind, beneath the surface of water and in the absence of air. Some estuary shells have been found in it, but few of pelagic origin, and no remains of those zoophytes and crinoidea so abundant in the mountain limestone and other rocks. Coal beds exist in Europe, Asia, and America, and have hitherto been esteemed as the most valuable of mineral productions, from the important services which the substance renders in manufactures and in domestic economy. It is to be remarked, that there are some local variations in the arrangement of coal beds. In France, they rest immediately on the granite and other primary rocks, the intermediate strata not having been found at those places. In other countries, traces of coal are found in the Devonian or Old Red Sandstone formation. These last circumstances may only show that different parts of the earth's surface did not all witness the same events of a certain fixed series exactly at the same time.

Some features of the condition of the earth during the deposition of the carboniferous group, are made out with a clearness which must satisfy most minds. First we are told of a time when carbonate of lime was formed in vast abundance along the shores and islands of the ocean, accompanied by an unusually large population of corals and encrinites; while in some parts of the earth there were pieces of dry land, covered with a luxuriant vegetation. Next we have a comparatively brief period of volcanic disturbance, (when the conglomerate was formed.) Then the causes favourable to the so abundant production of limestone, and the large population of marine radiates, decline, and we find the masses of dry land increase in number and extent, and begin to bear an amount of forest vegetation, far exceeding that of the most sheltered tropical spots of the present surface. The climate, even in the latitude of Baffin's Bay, was torrid; and the atmosphere has been supposed by some to have contained a larger charge of carbonic acid gas (the material of vegetation) than it now does. The forests or thickets of the period included no plants specifically the same with those now known upon earth. They mainly consisted of gigantic vegetables, many of which are not represented by any existing types, while others are akin to kinds which, in temperate climes at least, are now only found in small and lowly forms. That these forests grew upon a Polynesia, or multitude of small islands, is considered probable, from similar vegetation being now found in such situations within the tropics.

With regard to the circumstances under which the masses of vegetable matter were transformed into successive coal strata, geologists are divided. From examples seen at the present day, at the mouths of such rivers as the Mississippi, which traverse extensive sylvan regions, and from other circumstances to be adverted to, it is held likely by some that the vegetable matter, the rubbish of decayed forests, was carried by rivers into estuaries, and there accumulated in vast natural rafts, until it sunk to the bottom, where an overlayer of sand or mud would prepare it for becoming a stratum of coal. Others conceive that the vegetation first passed into the condition of a peat moss, that a subsidence then exposed it to be overrun by the sea, and covered with a layer of sand or mud; that a subsequent uprise made the mud dry land, and fitted it to bear a new forest, which afterwards, like its predecessor, became a bed of peat; that, in short, by repetitions of this process, the alternate layers of coal, sandstone, and shale, constituting the carboniferous group were formed. It is favourable to this last view that marine fossils are rarely found in the body of the coal itself, though abundant in the shale layers above and below it; also that in several places erect stems of trees are found with their roots still fixed in the shale beds, and crossing the sandstone beds at almost right angles, showing that these, at least, had not been drifted from their original situations. On the other hand, it is not easy to admit such repeated risings and sinkings of surface as would be required, on this hypothesis, to form a series of coal strata. Perhaps we may most safely rest at present with the supposition that coal has been formed under both classes of circumstances, though in the latter only as an exception to the former.

The plants of the carboniferous period have been investigated with great care, by several able naturalists, and above eight hundred species have been ascertained; a result most creditable to the inquirers, but which is far from satisfactory to the world, seeing that we have 80,000 living species, and cannot suppose the flora of that remote age to have been so much more limited. It must, however, be observed, that there are many conceivable circumstances to account for the non-preservation or transmission of many of the plants of this era. The numerous fungi, and other lowly forms, appear quite unlikely to have left clear memorials of themselves in the rocks, or in the masses of coal; and it has even been ascertained by experiment, that some of the highest forms of vegetation perish with surprising quickness in water. If we might assume, nevertheless, that the plants actually ascertained, form in any degree a representation of the flora of this period, they would imply that the early terrestrial botany of our globe was composed chiefly of plants of comparatively simple form and structure.*

In the ranks of the vegetable kingdom, the lowest place is taken by plants of cellular tissue, and which have no flowers, (*cryptogamia*,) as sea-weeds, lichens, mosses, fungi, ferns. Above these stand plants with vascular tissue, and bearing flowers, in which again there are two great subdivisions; first, plants having one seed-lobe, (*monocotyledons*,) and in which the new matter is added within, (the cane and palm are examples;) second, plants having two seed-lobes, (*dicotyledons*,) and in which the new matter is added on the outside under the bark, (the pine, elm, oak, and all the British forest-trees are examples;) these subdivisions also ranking in the order in which they are here stated. Now it is found that the predominant plants of the coal era are of the cellular and cryptogamic kind, while the dicotyledons are comparatively rare. There is, indeed, one exogenous family, which occurs in considerable numbers, and, perhaps, figured more conspicuously in the living woods than in the dead coal beds—namely, the conifers; but this, again, is held as the lowest family of its class.

* See Appendix, C.

That many trees of higher families now existed, seems unlikely, when we learn, that such trees occur in considerable numbers in subsequent formations, showing that there was nothing positively to forbid their being preserved in the coal measures, if they had then existed.

The master-form or type of the era was the *fern*, or breckan, of which about one hundred and thirty species have been ascertained as entering into the composition of coal. The ferns are plants which thrive best in warm, shaded, and moist situations. In tropical countries, where these con ditions abound, there are many more species than in temperate climes, and some of these are arborescent, or of a tree-like size and luxuriance.* The ferns of the coal strata have been of this magnitude, and that without regard to the regions of the earth where they are found. In the coal of Baffin's Bay, of Newcastle, and of the torrid zone, alike, are the fossil ferns arborescent, showing that, in that era, the present tropical temperature, or one even higher, existed in very high latitudes.

In the swamps and ditches of England there

* A specimen from Bengal, in the staircase of the British Museum, is forty-five feet high.

grows a plant called the horse-tail, (*equisetum*,) having a succulent, erect, jointed stem, with slender leaves, and a scaly catkin at the top. A second large section of the plants of the carboniferous era were of this kind, (*equisetaceæ*,) but, like the fern, reaching the magnitudes of trees. While existing equiseta rarely exceed three feet in height, and the stems are generally under half an inch in diameter, their kindred, entombed in the coal beds, seem to have been generally fourteen or fifteen feet high, with stems from six inches to a foot in thickness. It is to be remarked that plants of this kind (forming two genera, the most abundant of which is the *calamites*) are only represented on the present surface by plants of the same *family:* the *species* which flourished at this era gradually lessen in number as we advance upwards in the series of rocks, and disappear before we arrive at the tertiary formation.

The club-moss family (*lycopodiaceæ*) are other plants of the present surface, usually seen in a lowly and creeping form in temperate latitudes, but presenting species which rise to a greater magnitude within the tropics. Many specimens of this family are found in the coal beds; it is thought they have contributed more to the substance of the coal than any other family. But, like the ferns and equisetaceæ, they rise to a prodigious magnitude. The lepidodendra (so the fossil genus is called) have probably been from sixty-five to eighty feet in height, having at their base a diameter of about three feet, while their leaves measured twenty inches in length. In the forests of the coal era, the lepidodendra would enjoy the rank of firs in our forests, affording shade to the only less stately ferns and calamites. The internal structure of the stem, and the character of the seed-vessels, show them to have been a link between single-lobed and double-lobed plants—a fact worthy of note, as it favours the idea of a progress in vegetable creation, in conformity with advancing conditions. It is also curious to find a missing link of so much importance in a genus of plants which has long ceased to have a living place upon earth.

The other leading plants of the coal era are without representatives on the present surface, and their characters are in general less clearly ascertained. Amongst the most remarkable are—the *sigillaria*, of which large stems are very abundant, showing that the interior has been soft, and the exterior fluted, with separate leaves inserted in vertical rows along the flutings—and the *stigmaria*, a plant apparently calculated to flourish in marshes, or pools, having a short, thick, fleshy stem, with a dome-shaped top, from which sprung branches of from twenty to thirty feet long. Amongst monocotyledons were some palms, (*flabellaria* and *nœggerathia*,) besides a few not distinctly assignable to any class.

The conifers of the coal are comparatively rare, and are only as yet found in isolated cases, and in sandstone beds. One discovered in the Craigleith quarry, near Edinburgh, consisted of a stem about two feet thick, and forty-seven feet in length. Others were afterwards found, both in the same situation, and at Newcastle. Leaves and fruit being wanting, an ingenious mode of detecting the nature of these trees was devised by some naturalists residing in the northern capital.* Taking thin polished cross slices of the stem, and subjecting them to the microscope, they detected the structure of the wood to be that of a cone-bearing tree, by the presence of certain "reticulations" which distinguish that family, in addition to the usual radiating and concentric lines. That

* See Witham, on the Internal Structure of Fossil Vegetables. 1834.

particular tree was concluded to be an araucaria, a species now found in Norfolk Island, in the South Sea, and in a few other remote situations. The coniferæ of this era may be said to form the dawn of dicotyledonous trees, to which, it has already been noticed, the lepidodendra are a link from the monocotyledons. The concentric rings of the Craigleith and other coniferæ of this era have been mentioned. It is interesting to find in these a record of the changing seasons of those early ages, when as yet there were no human beings to observe time or tide. The rings are clearly traced; but it is observed that they are more slightly marked than is the case with their family at the present day, as if the changes of temperature had been within a narrower range.

Such (if we are to be allowed to rest with positive evidence) was the vegetation of the carbonigenous era, composed of forms low in the botanical scale, mostly flowerless and fruitless, but luxuriant and abundant beyond what the most favoured spots on earth can now show. The rigidity of the leaves of its plants, and the absence of fleshy fruits and farinaceous seeds, unfitted it to afford nutriment to animals; and, monotonous in its forms, and destitute of brilliant colouring, its sward probably unenlivened by any of the smaller flowering herbs, its shades uncheered by the music of birds, it must have been a sombre scene to a human visitant. But neither man nor any other animals were then in existence to look for such uses or such beauties in this vegetation. It was serving other and equally important ends, clearing perhaps the atmosphere of matter noxious to animal life, and certainly storing up mineral masses which were in long subsequent ages to prove of the greatest service to the human race, even to the extent of favouring the progress of its civilization.

Traces of land plants previous to the Carboniferous era are isolated at the best, and, till we know more about them, they cannot be allowed greatly to affect our views of the botanical history of the globe. Geologists speak of a fern leaf in the Silurians of Wales; in those of America, a plant apparently allied to the lepidodendron; in the American lower Old Red Sandstone, some allied to ferns. These phenomena, even if fully established, would not interfere with general deductions from the mass of early land vegetation found in the coal era. There might be small pieces of land bearing vegetation long before the existence of the masses which produced the great coal flora; and from such pieces of land might those early specimens have been wafted.

The Carboniferous formation exhibits a scanty zoology, compared either with those which go before, or those which come after. The mountain limestone, indeed, deposited at the commencement of it, abounds unusually in polypiaria, crinoidea, and mollusca; but when we ascend to the coal-beds themselves, the case is altered, and these marine remains altogether disappear. We have then only a limited variety of shell mollusks, with fragments of a few species of fishes, and these are rarely or never found in the coal seams, but in the shales alternating with them. Among the fishes, the conspicuous form is that Sauroid family which we have seen commence in the Old Red Sandstone. It receives its name in consequence of a character of teeth, scales, and even osteology, resembling that of the Sauria, and evidently leading on to that section of reptiles.* One of the most noted species is the *Megalichthys Hibbertii*, discovered by Dr. Hibbert Ware, in a limestone bed at Burdiehouse, near Edinburgh, and of which other specimens have been found in

* See Appendix, D.

the coal measures of Yorkshire, and low coal shales of Newcastle. The enormous size of the animal is inferred from teeth belonging to it, not less than four inches long. At this point we find the first traces of land animals, in the fossil remains of terrestrial insects,* and the foot-prints of reptiles, the first in England, the latter in America.

Coal strata are nearly confined to the group termed the carboniferous formation. Thin beds are not unknown afterwards, but they occur only as a rare exception. It is therefore thought that the most important of the conditions which allowed of so abundant a terrestrial vegetation,—whatever these were,—had ceased about the time when this formation was closed.

The termination of the carboniferous formation is marked by symptoms of volcanic violence, which

* "Two species [of insects], belonging to the family of Curculionidæ, have been found in the coal-fields of Coalbrook Dale, as well as a neuropterous insect, which closely resembles the genus Corydalis now living in Carolina ; also a libellula, or an insect related to the Phasmidæ. * * Count Sternberg has likewise announced the discovery of a fossil scorpion in the coal-measures at Chomle, near Radnitz, in Bohemia. It is easily conceivable that, as insects could only leave traces of their existence in exceptional and very rare instances, it is very improbable that we should ever have a satisfactory knowledge of this part of the fauna of the ancient formations."—*D'Archiac and De Verneuil on the Fossils of the Older Deposits, &c. Geol. Trans.* vi. (2*d ser.*) 330.

some geologists have considered to denote the close of one system of things and the beginning of another. Coal-beds generally lie in basins, as if following the curve of the bottom of seas. But there is no such basin which is not broken up into pieces, some of which have been tossed up on edge, others allowed to sink, causing the ends of strata to be in some instances many yards, and in a few, several hundred feet, removed from the corresponding ends of neighbouring fragments. These are held to be results of volcanic movements below, the operation of which is further seen in numerous upbursts and intrusions of fire-born rock, (trap.) That these disturbances took place about the close of the formation, and not later, is shown in the fact of the next higher group of strata being comparatively undisturbed. Other symptoms of this time of violence are seen in the beds of conglomerate which occur amongst the first strata above the coal. These, as usual, consist of fragments of the elder rocks, more or less worn from being tumbled about in agitated water, and laid down in a mud paste, afterwards hardened.* It is

* Volcanic disturbances break up the rocks; the pieces are worn in seas; and a deposit of conglomerate is the consequence. Of porphyry, there are some such pieces in the conglomerate of Devonshire, three or four tons in weight.

to be admitted for strict truth, that in some parts of Europe the carboniferous formation is followed by superior deposits, without the appearance of such disturbances between their respective periods; but apparently this case is exceptive. That disturbance was general, is supported by the further and important fact of the destruction of many forms of organic being previously flourishing, particularly of the vegetable kingdom.

ERA OF THE NEW RED SANDSTONE.

TERRESTRIAL ZOOLOGY COMMENCES WITH REPTILES.

FIRST TRACES OF BIRDS.

THE next volume of the rock series refers to an era distinguished by an event of no less importance than the abundant appearance of land animals. The *New Red Sandstone System* is subdivided into groups, some of which are wanting in some places. The lowest beds are those alluded to in the preceding chapter as presenting indications of disturbance. Next occur the strata of the Magnesian Limestone, denoting from their composition a recürrence of circumstances favourable to animal life.* Beds usually called the Upper

* The Lower New Red and the Magnesian Limestone have lately been called the Permian System, from their being largely developed in the ancient kingdom of Permia.

New Red Sandstone, followed in Germany by the Muschelkalk, or Shell Limestone, are next in the series; they are crowned by a group of Variegated Marls. This section of rocks forms, in whole, a sort of transitionary series, from the Carboniferous to the Oolitic; and, though peculiar in mineral constitution, might, as far as fossils are concerned, be very properly divided between the preceding and succeeding formations.

The plants of this era are few and unobtrusive. Equiseta, calamites, ferns, Voltzia, and a few of the other families, found so abundantly in the carboniferous series, here present themselves, but in diminished size and quantity.

The types of animal being which existed before —zoophytes, conchifers, mollusks, crustacea, and fishes—continue to appear in the New Red Sandstone Rocks, being most numerous in the limestone beds, but particularly in the German sub-group of the Muschelkalk. All of these great classes are aquatic. Hitherto, no distinct traces of land-walking or air-breathing animals, above the invertebrate grades, have been presented to our observation. But now, in this New Red formation, we discover the remains and other appearances of a group of vertebrate animals, most of which are fitted to breathe the atmosphere and tread the solid surface—namely, reptiles. These are soon presented to us in abundance of specimens and of species, and they continue for some subsequent ages to be the predominant tenants of the earth; insomuch that one long period has been emphatically called the Age of Reptiles. In that time, it will be found we have few traces of birds, and hardly any of the mammalian tribes.

The earliest traces of reptiles are few and scattered, so as to convey the idea that we have yet much to discover respecting the origin of the class upon earth. Here it must be remarked that the ingredients and arrangements of rocks, with fossil remains, do not form the sole materials of the history compiled by the geologist. He is equally contented when he can find an intelligible fact told by what may be called a writing of nature upon these stone tablets. So low as the bottom of the carboniferous system, slabs are found marked over a great extent of surface with that peculiar corrugation or wrinkling which the receding tide leaves upon a sandy beach when the sea is but slightly agitated; and not only are these ripple-marks, as they are called, found on the surfaces, but casts of them appear on the under sides of slabs lying above. The phenomena suggest the time when the sand, ultimately formed into these stone slabs, was part of the beach of a sea of the carbonigenous era; when, left wavy by one tide, it was covered over with a thin layer of fresh sand by the next, and so on, precisely as such circumstances might be expected to take place at the present day. Sandstone surfaces, ripple-marked, present themselves throughout the subsequent formations: in those of the New Red, at more than one place in England, they further bear impressions of rain-drops which have fallen upon them—the rain, of course, of the inconceivably remote age in which the sandstones were formed. In the Greensill sandstone, near Shrewsbury, it has even been possible to tell from what direction the shower came which impressed the sandy surface, the rims of the marks being somewhat raised on one side, exactly as might be expected from a slanting shower falling at this day upon one of our beaches. These facts have the same kind of interest as the season rings of the Craigleith conifers, speaking of the identity of the familiar processes of nature in those early ages with those of our own.

Hearing of memorials of this kind will prepare

the reader to learn that the earliest intelligence we have respecting land-walking animals consists, in great part, of their mere footsteps impressed on the wet sand or mud, which afterwards became rock. Let no one undervalue such testimony. The fidelity of an impression from a foot, as certifying by what or whose foot the impression was made, is acknowledged in judicial procedure; and often has this kind of evidence fixed the opinion of judge and jury, when every other had failed.

If we confine our attention, however, to geological researches in the eastern continent, we find the first traces of reptiles in actual fossils of the Magnesian Limestone. By Professor Owen, who has carefully examined them, they are said to be of the lacertilian or lizard order, (specifically called by him palæosaurs, thecodonts, monitors, etc.,) but for the most part of gigantic size, and differing from modern lizards in very remarkable characters of the vertebræ, teeth, and dermal plates. To them, as to all the reptiles of this and several subsequent great periods, belonged a fish-like form of the vertebral column, in as far as its bones were biconcave, or shaped like a double egg-cup, a peculiarity regarded by this eminent anatomist as probably fitting the animal for partially marine habits. And that the full importance of this peculiarity of the early reptiles may be appreciated, the reader must be made aware that modern reptiles have a ball-and-socket form of the vertebræ—that is, a convexity at the one side fitting into the hollow of the adjacent bone; but this form only when they are mature animals, for in the embryotic state of the crocodile and of the frog the form has been ascertained to be biconcave, which gradually changes as the animal approaches perfection. The teeth of these early lacertilia were also fixed to the jaws in the manner of fish teeth.

Ascending to the Upper New Red Sandstone, forming part of what has received the subordinate name of the *Triassic System*, we are introduced to a new lacertilian, presenting some remarkable characters, and named the *Rhynchosaurus*. From the few fragments of the animal which have been discovered, it would appear to have had a toothless head, resembling that of a bird, and enclosed in a bony sheath; also a hinder toe directed backwards, in which feature we also see an assimilation to the next higher vertebrate class. Footmarks attributed to this animal confirm the appearances presented by the extraordinary arrangement of its locomotive organs.

In the same beds occur a few bones, and a great number of footsteps, which Professor Owen has fixed as the double memorials of a group of animals, to which he has given (from the structure of their teeth) the name of Labyrinthidonts, and which he classes with the *Batrachia*,—that order of reptiles to which the frog and toad belong. Those who are accustomed to regard this as a group of generally small and insignificant animals, will be surprised to learn that the labyrinthidonts were of the size of a large hog. Their footmarks, discovered alike in America and the elder continent, "bear a singular resemblance to the impression that would be made by the palm and expanded fingers and thumb of the human hand." But it is evident that the fore extremities of the animal had been, like those of the kangaroo and some other genera, much smaller than the hinder, some specimens of which measure eight inches by five. These batrachia, like the lacertilia, present affinities to the fish class in their biconcave vertebræ and the formation and arrangement of the teeth. Their nostrils being also, like those of the Sauria, placed near the extremity of the head, indicate a partially marine habitat, such an arrangement being designed to enable the animal to breathe while nearly altogether sunk in the water.

Quarries of the Upper New Red also present an abundance of footmarks attributed to tortoises, thus pointing to the contemporaneous existence of a third order of reptiles, the *Chelonia*. The first examples were discovered by the Rev. Dr. Duncan in the quarry of Corncockle Muir, Dumfries-shire, where the slabs incline at an angle of thirty-eight degrees, and the footmarks are distinctly traced up and down the slope, as if the animal had had occasion to pass in that direction only, possibly in its daily visits to the sea. Some slabs similarly impressed, in the Stourton quarries, Cheshire, are further marked with a shower of rain which we know to have fallen *afterwards*, for its little hollows are impressed in the footmarks also, though more slightly than on the rest of the surface, the comparative hardness of a trodden place having apparently prevented so deep an impression being made.

Above the lower beds of the Upper New Red, on the Continent, there exists a series which are hardly traceable in England, the celebrated *Muschelkalk*, and here, for the first time, do we find examples of a group of reptiles which have excited more attention than perhaps any other. The same group, it may be remarked, occurs in the English lias and subsequent formations; but the mere fact of writing in England should not make us postpone to that place an order of beings which we find earlier in another portion of what, geologically, may be regarded as but one great zoological province. These animals, called collectively, *Enaliosauria*, or *Marine Saurians*, abounded throughout a long period of the earth's history, while mammalian life was yet hardly developed, but they disappeared in what we shall have to speak of as the Cretaceous Era, and since then have hardly even been represented upon our globe. The *Ichthyosaur*, of which ten species have been distinguished, was an animal of marine habits and great bulk, (reaching about thirty feet in length,) in which to the form of the fish there were united, in a remarkable way, characters of animals higher in the scale. A body, framed upon a purely piscine vertebral column, containing a huge voracious stomach, and terminating in a vertically expanded tail, in which respect it also preserved the fish character, was furnished with the head of a crocodile, and four fins approximating to the character of the paddles of the turtle. Over all this was a skin resembling that of the cetaceous animals. Nor should it be omitted that the sternum or breast-bone presents a structure resembling that of the ornithorhynchus or duck-rat of Australia. The vast jaws of this animal, having a stretch of seven feet; its eye resting in a socket eighteen inches in diameter, and defended by an apparatus of bony plates, like that of a bird of prey; the powerful range of teeth, and the position of the breathing apertures near the extremity of the snout; all speak to the naturalist of ferocious habits like those of the modern crocodile, to which the ichthyosaur may be considered as a link from the predaceous fish. A curious light has been thrown upon these habits by the pellets ejected from the stomach of the animal, which have been found in great quantities in a fossilized state, and bear the name of *coprolites.* There we find fragments not only of fish, but of reptiles, arguing that the animal must have been a destructive creature both to its own class and to that below it.

The genus next in importance is the *Plesiosaurus*, so called as being near to the saurian character. This animal was under eighteen feet long, and altogether a feebler creature than the Ichthyosaur, which seems to have made it a prey. Yet it was itself one of the destructive potentates of the early seas. A body, generally fish-like, though framed on vertebræ presenting less concave sides, and which terminated in a short tail, serving only as a rudder, was furnished with a long neck and small head, together with four slender paddles, more cetacean than those of the Ichthyosaur. Moving, like that animal, quickly in the water, by means of the special organs designed for the purpose, the Plesiosaur would have a further advantage in its long, flexible, serpent-like neck; but the small size of the head, in which we find some lacertian peculiarities united with characters mainly saurian, must have rendered it a much less formidable creature than that last described. Professor Owen regards it as fitted to live near shores and to ascend estuaries.

Of another enaliosaurian species, the Nothosaurus, we may only remark that it presents characters suggesting an approach to the crocodilians.

The different degrees in which we find animal life developed in different regions of the present surface, prepare us to hear that there are, in America, appearances of animals having lived at this time, somewhat superior to any of the types found in our hemisphere. The attention of the geologists of the United States has been called to certain footmarks in the sandstones of the valley of Connecticut, indicative of birds of the orders *Grallatores* (waders) and *Rasores* (scrapers.) "The footsteps appear in regular succession on the continuous track of an animal, in the act of walking or running, with the right and left foot always in their relative places. The distance of the intervals between each footstep on the same track is occasionally varied, but to no greater amount than may be explained by the bird having altered its pace. Many tracks of different individuals and different species are often found crossing each other, and crowded, like impressions of feet upon the shores of a muddy stream, where ducks and geese resort."* Some of these prints indicate small animals, but others denote birds of what would now be an unusually large size, one having a foot fifteen inches in length, and a stride of from four to six feet. There are anomalies in the forms of some of the feet; but of their being the vestiges of birds no doubt seems any longer to exist. There is still, however, an uncertainty re-

* Dr. Buckland, quoting an article by Professor Hitchcock, in the American Journal of Science and Arts, 1836.

garding the date of the rocks which present these memorials, for the phenomena of superposition only denote their being between the carboniferous and cretaceous formations, and an exact place is assigned them, merely upon the strength of the discovery that they present fish of certain genera never found above the Triassic series. Along with distinctly ornithic footmarks are those of the Labyrinthidont. Altogether above thirty species of Triassic birds are made out from these vestiges by the American geologists.*

* Early in 1845, the discovery of footmarks, apparently of wading birds and the batrachian reptiles, was announced as having taken place in Westmoreland County, Pensylvania, pretty far down in the carboniferous series.

ERA OF THE OOLITE.

COMMENCEMENT OF MAMMALIA.

THE chronicles of this period consist of a series of beds, mostly calcareous, taking their general name (*Oolite system*) from a conspicuous member of them—the oolite—a limestone composed of an aggregation of small round grains or spherules, and so called from its fancied resemblance to a cluster of eggs, or the roe of a fish. This texture of stone is novel and striking. It is supposed to be of chemical origin, each spherule being an aggregation of particles round a central nucleus. The oolite system is largely developed in England, France, Westphalia, and Northern Italy; it appears in Northern India and Africa, and patches of it exist in Scotland, and in the vale of the Mississippi. It may of course be yet discovered in many other parts of the world.

The series, as shown in the neighbourhood of Bath, is (beginning with the lowest) as follows: 1. Lias, a set of strata variously composed of limestone, clay, marl, and shale, clay being predominant; 2. Lower oolitic formation, including, besides the great oolite bed of central England, fullers'-earth beds, forest marble, and cornbrash; 3. Middle oolitic formation, composed of two subgroups, the Oxford clay and coral rag, the latter being a mere layer of the works of the coral polype; 4. Upper oolitic formation, including what are called Kimmeridge clay and Portland oolite. In Yorkshire there is an additional group above the lias, and in Sutherlandshire there is another group above that again. In the wealds (moorlands) of Kent and Sussex, there is, in like manner, above the fourth of the Bath series, another additional group, to which the name of the *Wealden* has been given, from its topographical situation, and which, composed of sandstones and clays, is subdivided into Purbeck beds, Hastings sand, and Weald clay.

There are no particular appearances of disturbance between the close of the new red sandstone and the beginning of the oolite system, as far as has been observed in England. Yet there is a great change in the materials of the rocks of the two formations, showing that, while the bottoms of the seas of the one period had been chiefly arenaceous, those of the other were chiefly clayey and limy. And there is an equal difference between the two periods in respect of both botany and zoology. While the new red sandstone shows comparatively scanty traces of organic creation, those in the oolite are extremely abundant, particularly in the department of animals, and more particularly still of sea mollusca. Geologists describe the animals of the oolitic system as different in species from those of the preceding age, and also from those which succeed; but there are in reality no certain marks establishing distinction of species, and here, as in similar cases, we are only to understand that the animals display certain external peculiarities. The distinguishing characters, such as they are, appear to be uniform over a great space. "In the equivalent deposits in the Himalayan Mountains, at Fernando Po, in the region north of the Cape of Good Hope, and in the Run of Cutch, and other parts of Hindostan, fossils have been discovered, which, as far as English naturalists who have seen them can determine, are undistinguishable from certain oolite and lias fossils of Europe."*

The dry land of this age presented cycadeæ, "a beautiful class of plants between the palms and conifers, having a tall, straight trunk, terminating in a magnificent crown of foliage."† There were tree ferns, but in smaller proportion than in former ages; also equisetaceæ, lilia, and coniferæ. The vegetation was generally analogous to that of the Cape of Good Hope and Australia, which seems to argue a climate between the tropical and temperate. It was, however, sufficiently luxuriant, in some instances, to produce thin seams of coal, for there are such in the oolite formation of both Yorkshire and Sutherland. The sea, as for ages before, contained algæ, of which, however, only a few species have been preserved to our day. The lower classes of the inhabitants of the ocean were unprecedentedly abundant, the polypiaria forming whole strata of themselves. The crinoidea and

* Murchison's Silurian System, p. 583.

† Buckland.

echinites were also extremely numerous. Shell mollusks, in hundreds of new species, occupied the bottoms of the seas of those ages, while of the swimming and piratical molluscs, the ammonites and belemnites, there were also many scores of varieties. The belemnite here calls for some particular notice. It commences in the oolite, and terminates in the next formation. It is an elongated, conical shell, terminating in a point, and having, at the larger end, a cavity for the residence of the animal, with a series of air-chambers below. The animal, placed in the upper cavity, could raise or depress itself in the water at pleasure by a pneumatic operation upon the air tube pervading its shell. Its tentacula, sent abroad over the summit of the shell, searched the sea for prey. The creature had an ink bag with which it could muddle the water around it, to protect itself from more powerful animals, and strange to say, this has been found so well preserved, that an artist has used it in one instance as a pigment, wherewith to delineate the belemnite itself.

The crustacea discovered in this formation are less numerous. There are many fishes, some of which (*acrodus*, *psammodus*, &c.) are presumed, from remains of their palatal bones, to have been of a gigantic cartilaginous class, (*placoidean,*) now represented by such as the crestaceon. It has been considered by Professor Owen as worthy of notice, that, the crestaceon being an inhabitant of the Australian seas, we have, in both the botany and ichthyology of this period, an analogy to that Continent. The pycnodontes, (thick toothed,) and lepidoides, (having thick scales,) are other families described by M. Agassiz as extensively prevalent.

In the English lias there is a vast abundance of the enaliosauria which we have seen commence in the foreign Muschelkalk, and, in addition to these, specimens of *Pterosauria* or *Winged Saurians,* a type of being, the most new, perhaps, of all which the geological record has presented to us. The Pterodactyls, as the animals of this order have been called, were saurians of small size compared with their associates, being not larger than a modern cormorant; but the marvel in their case consists in bat-like wings extended upon the fore-finger, by which the animal was enabled to pursue its way in the air. This order became extinct in the time of the chalk formation. The only existing animal of which it may even remind us is the *draco volans* or flying lizard, which has a membrane by which to support itself in leaping from tree to tree.

In the proper oolite, there is added an enaliosaurian (the *Pliosaur*) in which there is a very close approach made to the crocodilian order, but upon a scale of enormous magnitude, the animal having apparently been as large as the existing whales. Here, too, we find the true *crocodilia* largely developed, and five genera have been described (*Teleosaurus, Steneosaurus, Cetiosaurus*, &c.) The two first are like crocodiles of our own time in all respects, but a somewhat greater bulk, and certain peculiarities, indicating more aquatic habits. The last derives its name from the approximation to the whale tribes seen in the form of its vertebræ. In this group there is a genus presenting ball-and-socket vertebræ, and thus proving its advanced character, but, strange to say, the concavity is in this case directed backwards, instead of forwards, which is the universal arrangement in similar cases, in our era.

The first glimpse of the highest class of the vertebrate sub-kingdom—*Mammalia*—is obtained from the Stonesfield slate, where there have been found several specimens of the lower jaw-bone of a quadruped evidently insectivorous, and inferred, from peculiarities of structure, to have belonged to the marsupial family, (pouched animals.)* It may be observed, although no specimens of so high a class of animals as mammalia are found earlier, such may nevertheless have existed: the defect may be in our not having found them; but, other things considered, the probability is that heretofore there were no mammifers. It is an interesting circumstance that the first mammifers found should have belonged to the marsupialia, when the place of that order in the scale of creation is considered. In the imperfect structure of their brain, deficient in the organs connecting the two hemispheres—and in the mode of gestation, which is only in small part uterine—this family is usually regarded as only a little advanced above the character of the bird.

The highest part of the oolitic formation presents some phenomena of an unusual and interesting character, which demand special notice.

* Fragments attributed to a cetaceous animal, another humble form of the mammal class, have likewise been found in the great oolite, near Oxford.

Immediately above the upper oolitic group in Buckinghamshire, in the vicinity of Weymouth, and other situations, there is a thin stratum, usually called by workmen the *dirt-bed*, which appears, from incontestable evidence, to have been a soil, formed, like soils of the present day, in the course of time, upon a surface which had previously been the bottom of the sea. The dirt-bed contains exuviæ of tropical trees, accumulated through time, as the forest shed its honours on the spot where it grew, and became itself decayed. Near Weymouth there is a piece of this stratum, in which stumps of trees remain rooted, mostly erect or slightly inclined, and from one to three feet high; while trunks of the same forest, also silicified, lie imbedded on the surface of the soil in which they grew.

Above this bed lie those which have been called the Wealden, from their full development in the Weald of Sussex; and these as incontestably argue that the dry land forming the dirt-bed had next afterwards become the area of brackish estuaries or lakes partially connected with the sea; for the Wealden strata contain exuviæ of freshwater tribes, besides those of the great saurians and chelonia. The area of this estuary comprehends the whole south-east province of England. A geologist thus confidently narrates the subsequent events: "Much calcareous matter was first deposited [in this estuary], and in it were entombed myriads of shells, apparently analogous to those of the vivipara. Then came a thick envelope of sand, sometimes interstratified with mud; and, finally, muddy matter prevailed. The solid surface beneath the waters would appear to have suffered a long continued and gradual depression, which was as gradually filled, or nearly so, with transported matter; in the end, however, after a depression of several hundred feet, the sea again entered upon the area, not suddenly or violently —for the Wealden rocks pass gradually into the superincumbent cretaceous series—but so quietly, that the mud containing the remains of terrestrial and fresh-water creatures was tranquilly covered up by sands replete with marine exuviæ."* A subsequent depression of the same area, to the depth of at least three hundred fathoms, is believed to have taken place, to admit of the deposition of the cretaceous beds lying above.

* De la Beche's Geological Researches, p. 344.

From the scattered way in which remains of the larger terrestrial animals occur in the Wealden, and the intermixture of pebbles of the special appearance of those worn in rivers, it is also inferred that the estuary which once covered the south-east part of England was the mouth of a river of that far-descending class of which the Mississippi and Amazon are examples. What part of the earth's surface presented the dry land through which that and other similar rivers flowed, no one can tell. It has been surmised, that the particular one here spoken of may have flowed from a point not nearer than the site of the present Newfoundland. Professor Phillips has suggested, from the analogy of the mineral composition, that anciently elevated coal strata may have composed the dry land from which the sandy matters of these strata were washed. Such a deposit as the Wealden almost necessarily implies a local, not a general condition; yet it has been thought that similar strata and remains exist in the Pays de Bray, near Beauvais. This leads to the supposition that there may have been, in that age, a series of river-receiving estuaries along the border of some such great ocean as the Atlantic, of which that of modern Sussex is only an example.

The zoology of the Wealden is chiefly remarkable for the additions which it makes to the list of reptiles presented in previous formations. Besides some new crocodilia (*Suchosaurus* and *Goniopholis*), and several chelonia (*Tetrosternon*, etc.), we have here the principal constituents of a group, which Professor Owen has described as a distinct order, under the name of *Dinosauria*, the remaining form being the *Megalosaurus* of the oolite. These were terrestrial crocodile-like animals, with some features of organization recalling the lacertilia, and also such a massive and stately form of the extremities, as to remind us of the large land pachyderms. The animal last named, from twenty-five to thirty feet long, with an enormous muzzle furnished with strong teeth, must have been by far the most formidable land creature of its age. The very opposite habits of the Iguanodon, an equally huge herbivorous reptile, lead me to suspect an error in the classification; but—passing from this—its size and stately limbs are such as equally to excite our surprise. From the scapula or blade-bone of the remaining genus, the *Hylæosaurus*, the approximation of the whole of the dinosaurs to the mammalian type of structure has been inferred.

The first fossils referred to birds occur in the Wealden. They belonged to the wading order, and probably to the heron family. It has been thought, that the immediate connexion of these beds with land might account for their containing a terrestrial organic relic which the marine beds above and below did not possess.

ERA OF THE CRETACEOUS FORMATION.

The record of this period consists of a series of strata, in which chalk beds make a conspicuous appearance, and which is therefore called the cretaceous system or formation. In England, a long stripe, extending from Yorkshire to Kent, presents the cretaceous beds upon the surface, generally lying conformably upon the oolite, and in many instances rising into bold escarpments towards the west. The celebrated cliffs of Dover are of this formation. It extends into Northern France, and thence north-westward into Germany, whence it is traced into Scandinavia and Russia. The same system exists in North America, and probably in other parts of the earth not yet geologically investigated. Being a marine deposit, it establishes that seas existed at the time of its formation on the tracts occupied by it, while some of its organic remains prove that, in the neighbourhood of those seas, there were tracts of dry land.

The cretaceous formation in England presents beds chiefly sandy in the lowest part, chiefly clayey in the middle, and chiefly of chalk in the upper part, the chalk beds being never absent, which some of the lower are in several places. In the vale of the Mississippi, again, the true chalk is wholly, or all but wholly absent. In the south of England, the lower beds are (reckoning from the lowest upwards), 1. *Shankland* or *greensand*, "a triple alternation of sands and sandstones with clay;" 2. *Galt*, "a stiff blue or black clay, abounding in shells, which frequently possess a pearly lustre;" 3. *Hard* chalk; 4. Chalk with flints; these two last being generally white, but in some districts red, and in others yellow. The whole are, in England, about 1200 feet thick, showing the considerable depths of the ocean in which the deposits were made.

Chalk is a carbonate of lime, and the manner of its production in such vast quantities was long a subject of speculation among geologists. Some light seemed to be thrown upon the subject a few years ago, when it was observed, that the detritus of coral reefs in the present tropical seas gave a powder, undistinguishable, when dried, from ordinary chalk. It then appeared likely that the chalk beds were the detritus of the corals which lived in the oceans of that era. Mr. Darwin, who made some curious inquiries on this point, further suggested, that the matter might have intermediately passed through the bodies of worms and fish, such as feed on the corals of the present day, and in whose stomachs he has found impure chalk. This, however, cannot be a full explanation of the production of chalk, if we admit some more recent discoveries of Professor Ehrenberg. That master of microscopic investigation announces, that chalk is composed partly of "inorganic particles of irregular elliptical structure and granular slaty disposition," and partly of shells of inconceivable minuteness, "varying from the one-twelfth to the two hundred and eighty-eighth part of a line"—a cubic inch of the substance containing above ten millions of them! The chalk of the north of Europe contains, he says, a large proportion of the inorganic matter; that of the south, a larger proportion of the organic matter, being in some instances almost entirely composed of it. He has been able to classify many of these creatures, some of them being allied to the nautili, nummuli, cyprides, &c. The shells of some are calcareous, of others siliceous. M. Ehrenberg has likewise detected microscopic sea-plants in the chalk.

The distinctive feature of the uppermost chalk beds in England is the presence of flint nodules. These are generally disposed in layers parallel to each other. It was readily presumed by geologists that these masses were formed by a chemical aggregation of particles of silica, originally held in solution in the mass of the chalk. But whence the silica in a substance so different from it? Ehrenberg suggests that it is composed of the siliceous coverings of a portion of the microscopic creatures, whose shells he has in other instances detected in their original condition. It is remarkable that the chalk *with* flint abounds in the north of Europe; that *without* flints in the south; while in the northern chalk siliceous animalcules are wanting, and in the southern present in great quantities. The conclusion seems natural, that in the one case the siliceous exuviæ have been left in their original form; in the other, dissolved chemically, and aggregated on the common principle of chemical affinity into nodules of flint, probably concentrating, in every instance, upon a piece of decaying organic matter, as has been the case with the nodules of ironstone in the earlier rocks, and the spherules of the oolite.

What is more remarkable, M. Ehrenberg has ascertained that at least fifty-seven species of the microscopic animals of the chalk, being infusoria and calcareous-shelled polythalamia, are still found living in various parts of the earth. These species are the most abundant in the rock. Singly they are the most unimportant of all animals, but in the mass, forming as they do such enormous strata over a large part of the earth's surface, they have an importance greatly exceeding that of the largest and noblest of the beasts of the field. Moreover, these species have a peculiar interest, as the only specific types of that early age which have survived to the present day. Species of sea mollusks, of reptiles, and of mammifers, have been changed again and again, since the cretaceous era; and it is not till a long subsequent age that we find the first traces of any other of even the humblest species which now exist; but here have these humble infusoria and polythalamia kept their place on earth through all its revolutions since that time,—are we to say, persistent through a continuing uniformity in the conditions under which they have lived, while all other animals have been exposed to circumstances productive of change?

All the ordinary and more observable orders of the inhabitants of the sea, except the cetacea, have been found in the cretaceous formation—zoophytes, radiaria, mollusks, crustacea, (in great variety of species,) and fishes in smaller variety. Down to this period, the placoid and ganoid fishes had, as far as we have evidence, flourished alone; now they decline, and we begin to find in their place fishes of two orders of superior organization, those which predominate in the present creation. These are osseous in internal structure, with corneous scales, the latter being circular in the one case, and pectinated or indented at one side in the other; hence the two orders are called respectively cycloid and ctenoid by M. Agassiz, who, as has been remarked, asserts that the outer covering of fishes is a sufficient indication of their whole structure. In Europe, no remains of the marine saurians have been found; they may be presumed to have become extinct in that part of the globe before this time. In America, however, remains of the plesiosaurus have been discovered in this part of the stratified series. The reptiles, too, so numerous in the two preceding periods, appear to have now much diminished in numbers. One of the most remarkable was the mosæsaurus, which seems to have held an intermediate place between the monitor and iguana, and to have been about twenty-five feet long, with a tail calculated to assist it powerfully in swimming.

Fuci abounded in the seas of this era, and confervæ are found enclosed in flints. Of terrestrial vegetation, as of terrestrial animals, the specimens in the European area are comparatively rare, rendering it probable that there was little dry land near. The remains are chiefly of ferns, conifers, and cycadeæ, but in the two former cases we have only cones and leaves. There have been discovered many pieces of wood containing holes drilled by the teredo, and thus showing that they had been long drifted about in the ocean before being entombed at the bottom.

The series in America corresponding to this, entitled the Ferruginous Sand formation, presents fossils generally identical with those of Europe, not excepting the fragments of drilled wood; showing that, in this, as in earlier ages, there was a parity of conditions for animal life over a vast tract of the earth's surface. To European reptiles, the American formation adds a gigantic one, styled the Saurodon, from the lizard-like character of its teeth.

We have seen that footsteps of birds are discovered in America, in the new red sandstone. Some similar isolated phenomena occur in the subsequent formations. In the slate of Glaris, in Switzerland, corresponding to the English galt, in the chalk formation, the remains of a bird have been found. From a chalk bed near Maidstone, have likewise been extracted some remains of a bird, supposed to have been of the long-winged swimmer family, and equal in size to the albatross.

ERA OF THE TERTIARY FORMATION—MAMMALIA ABUNDANT.

The chalk-beds are the highest which extend over a considerable space; but in hollows of these beds, comparatively limited in extent, there have been formed series of strata—clays, limestones, marls, alternating—to which the name of the *Tertiary Formation* has been applied. London and Paris alike rest on basins of this formation, and another such basin extends from near Winchester, under Southampton, and re-appears in the Isle of Wight. A stripe of it appears along the east coast of North America, from Massachusetts to Florida. It is also found in Sicily and Italy, insensibly blended with formations still in progress. Though comparatively a local formation, it is not of the less importance as a record of the condition of the earth during a certain period.

The hollows filled by the tertiary formation must be considered as the beds of estuaries left at the conclusion of the cretaceous period. We have seen that an estuary, either by the drifting up of its mouth, or a change of level in that quarter, may be supposed to have become an inland sheet of water, and that, by another change of the reverse kind, it may be supposed to have become an estuary again. Such changes the Paris basin appears to have undergone oftener than once, for, first, we have there a fresh-water formation of clay and limestone beds; then, a marine-limestone formation; next, a second fresh-water formation, in which the material of the celebrated *plaster of Paris* (gypsum) is included; then a second marine formation of sandy and limy beds; and finally, a third series of fresh-water strata. Such alternations occur in other examples of the tertiary formation likewise.

Between the close of the chalk age and the beginning of the tertiary, a greater gap occurs in the fossil history of the earth than at any other period. The species now presented are almost wholly new, as if a considerable time had elapsed, during which the usual progressive change of animals had been going on, but, from geognostic causes, without the usual record having been kept. From this point, too, as we ascend in the series, we find more and more species identical with those still existing upon earth, as if we had now reached the dawn of the present state of the zoology of our planet. By the study of the shells alone, Mr. Lyell has formed a division of the whole term into four sub-periods, to which he has given names with reference to the proportions which they respectively present of surviving species—first, eocene; second, miocene; third, older pliocene; fourth, newer pliocene. This division, however, is to be regarded as not safely applicable to the Tertiaries generally, except as a convenient means of indicating various portions of the series.

The eocene period presents, in three continental groups, 1238 species of shells, of which forty-two, or 3·5 per cent. yet flourish. Some of these are remarkable enough; but they all sink into insignificance beside the mammalian remains which the lower eocene deposits of the Paris basin present to us, showing that the land had now become the theatre of an extensive creation of the highest class of animals. Cuvier ascertained about fifty species of these, all of them long since extinct. About four-fifths are of the order *Pachydermata*, thick-skinned animals, to which our modern elephant, rhinoceros, horse, and pig belong. Nearly the whole of these, however, belong to a family which is now confined to South America and Sumatra, namely, the tapirs,—an animal of squat figure, and possessing a short proboscis, an inhabitant of the woods, and an herbivore, but of unsocial habits. It is curious to find that a family now so limited in its range, had formerly been distributed over France, England, and other parts of the earth. Naturalists have conferred the names, Palæotherium, Lophiodon, Coryphodon, &c., upon the ancient extinct tapirs, which seem chiefly to differ from modern species in a few peculiarities of the constitution of the teeth, and in having three, instead of four toes upon the fore feet. One British specimen seems to have been about a third larger than the modern animal.

Another section of the Paris eocene remains have served to reconstruct a family to which the general name *Anoplotherium* has been given, from a regard to its deficiency of all offensive or defensive weapons. These were the first examples of hooved animals as yet discovered upon earth; they were strictly herbivorous, and make a slight approach to the cervine or deer tribes. The common anoplothere was about the size of an ass, but less elevated from the ground, and with a tail of above three feet in length; it is supposed to have been of aquatic habits, and an expert swimmer and diver, but also given to browsing upon land. Associated with these we find the first example (chœropotamus) of an animal approaching to the hog tribe, being nearest to the peccary of South America.

We learn from the remainder of the Paris fossils, and from others found in the eocene that the earth now possessed fresh-water reptiles; serpents of the size of the boa; natatorial, wading, and rapacious birds; rodents (dormouse and squirrel); species allied to the racoon, the genette, and fox; also bats and monkeys. Lastly, the oldest tertiaries of America present us with the *Zeuglodon*, a herbivorous whale resembling the dugong, having a stinted development of the extremities, but an enormous tail, and reaching altogether the length of a hundred feet.

In the miocene sub-period, the shells give eighteen per cent. of existing species, showing a considerable advance from the preceding era with regard to the inhabitants of the sea. The advance in land animals is less marked, but yet considerable. The predominating forms are still pachyderms, and the tapiroid animals continue to be conspicuous. Here occur remains of the *Dinotherium*, a creature said to exhibit an affinity to the cetacea in the form of its head, and to the tapir in the character of its teeth. It is most distinguished by its huge size, being not less than eighteen feet long; it had a mole-like form of the shoulder-blade, conferring the power of digging for food, and a couple of tusks turning down from the lower jaw, by which it could have attached itself, like the walrus, to a shore or bank, while its body floated in the water. Dr. Buckland considers this and some similar miocene animals, as adapted to a semi-aquatic life, in a region where lakes abounded. Besides the tapirs, we have in this era animals allied to the glutton, the bear, the dog, the horse, the hog, and lastly, several felinæ, (creatures of which the lion is the type;) all of which are new forms, as far as we know. There was also an abundance of marine mammalia, seals, dolphins, lamantins, walruses, and whales.

The shells of the older pliocene give from thirty-five to fifty; those of the newer, from ninety to ninety-five per cent. of existing species. The pachydermata of the preceding era now disappear;

but others enter upon the scene—elephantoid animals, the hippopotamus, rhinoceros, and horse. All of these bear a striking resemblance to pachyderms of the same families still existing. We have, in the mastodon and mammoth, which succeed each other in the strata, elephants variously distinguished from the present by peculiarities in their dentition, and hence considered as of different species, though this is a purely arbitrary distinction. What is remarkable of these ancient animals is their having lived in countries so far beyond the present range of their family, namely, throughout the whole temperate region of Asia and Europe, (England not being excepted,) and even so far north as the seventieth degree of latitude. The mammoth also inhabited North America. Its chief external peculiarity was a pair of long curved tusks extending forwards and upwards from the upper jaw. The numerous remains of the animal in the most superficial strata, and the discovery (in 1801) of a specimen with its flesh and hide entire in a mass of ice at the mouth of the Lena in Siberia, show that it must have lived down to comparatively modern times.

The pliocene gives many other new families. From remains which have been found, however fragmentary in many cases, there cannot be a doubt that all the principal mammalian forms, except the highest, now existed throughout the earth, and in species which only differed from those now living in slight peculiarities, chiefly of dentition. Bears, badgers, hyænas, and feline animals; moles and other insectivores; otters and weasels; the wolf and dog, then roamed for prey as now; besides an extinct felina, the machairodus, possessing teeth like curved saws. England had beavers and bears, little different from living species; only, one of the former family was of huge bulk. We also had the hippopotamus and rhinoceros. Oxen, deer, camels, etc., now inhabited the great zoological province with which we are connected; and monkeys and apes passed far beyond the tropical regions to which they are now confined. In India, besides the pachyderms of the European eocene, there were ruminants in abundance (including an extraordinary one, of huge bulk, named the Sivatherium), carnivores, rodents, and insectivores. Here also were monkeys, of unusual bulk; but the most wonderful animal as yet discovered in this region was a tortoise, not distinguishable in any point of structure from a land species now living, but reaching the surprising length of eighteen feet. The discoveries among the tertiaries of South America have been of a not less interesting character, in as far as they equally show an approach to the existing zoological characters of that region. Dr. Lund, a Danish naturalist, presents us with a monkey, indicating the features of the platyrrhine or New World group; and the edentate order, which is still more peculiar to that region, is there preceded by examples of vast size. In the megatherium, megalonyx, scelidotherium, and mylodon, we have a family of sloths, of elephantine magnitude, which lived by breaking down and eating trees. The toxodon surprises us not less, being an equally huge member of the rodent order,—that order which now includes most of the smallest quadrupeds.

One remarkable circumstance connected with the tertiary formation remains to be noticed,—the prevalence of volcanic action at that era. In Auvergne, in Catalonia, near Venice, and in the vicinity of Rome and Naples, lavas exactly resembling the produce of existing volcanoes, are associated and intermixed with the lacustrine as well as marine tertiaries. The superficies of tertiaries in England is disturbed by two great swells, forming what are called anticlinal axes, one of which divides the London from the Hampshire basin, while the other passes through the Isle of Wight, both throwing the strata down at a violent inclination towards the north, as if the subterranean disturbing force had *waved* forward in that direction. The Pyrenees, too, and Alps, have both undergone elevation since the deposition of the tertiaries; and in Sicily there are mountains which have risen three thousand feet since the deposition of some of the most recent of these rocks. The general effect of these operations was of course to extend the land surface, and to increase the variety of its features, thus improving the natural drainage, and generally adapting the earth for the reception of higher classes of animals.

ERA OF THE SUPERFICIAL FORMATIONS.

COMMENCEMENT OF PRESENT SPECIES.

We have now completed our survey of the series of stratified rocks, and traced in their fossils the progress of organic creation down to a time which seems not long antecedent to the appearance of man. There are, nevertheless, memorials of still another era or space of time which it is all but certain did also precede that event.

The first that calls for notice is the phenomenon to which geologists have applied the term denudation. Great hitches and slips are detected in superficial strata,—such as, if left in their original state, must have caused considerable inequalities on the face of the country; yet all is found as smooth—the joinings are all as much reduced to a common level—as if some gigantic artificial force had been used for the purpose. Again, a great valley has been scooped out in the midst of sedimentary strata, leaving the edges of these facing each other from the opposite sides, with perhaps here and there an isolated mass starting up to the height of the two sides, being composed of matter which has resisted the agency by which the adjoining matter was removed. Here, it is thought, we see incontestable traces of the operation of moving water. The second fact we are called to notice is that, over the rock formations of all eras, in various parts of the globe, but confined in general to situations not very elevated, there is a layer of stiff clay, mostly of a blue colour, mingled with fragments of rock of all sizes, travel-worn, and otherwise, and to which geologists give the name of Diluvium, as being apparently the produce of some vast flood, or of the sea thrown into an unusual agitation. It seems to indicate that, at the time when it was laid down, much of the present dry land was under the ocean, a supposition which we shall see supported by other evidence. The included masses of rock have been carefully inspected in many places, and traced to particular parent beds at considerable distances. Connected with these phenomena are certain rock surfaces on the slopes of hills and elsewhere, which exhibit groovings and scratchings, such as we might suppose would be produced by a quantity of loose blocks hurried along over them by a flood. Another associated phenomenon is that called *crag and tail*, which exists in many places,—namely, a rocky mountain, or lesser elevation, presenting on one side the naked rock in a more or less abrupt form, and on the other a gentle slope; the sites of Windsor, Edinburgh, and Stirling, with their respective castles, are specimens of crag and tail. Finally, I may advert to certain long ridges of clay and gravel which arrest the attention of travellers on the surface of Sweden and Finland, and which are also found in the United States, where, indeed, the whole of these phenomena have been observed over a large surface, as well as in Europe. It is very remarkable that the direction from which the diluvial blocks have generally come, the lines of the grooved rock surfaces, the direction of the crag and tail eminences, and that of the clay and gravel ridges—phenomena, be it observed, extending over the northern parts of both Europe and America—*are all from the north and north-west towards the south-east.* We thus acquire the idea of a powerful current moving in a direction from north-west to south-east, carrying, besides mud, masses of rock which furrowed the solid surfaces as they passed along, abrading the north-west faces of many hills, but leaving the slopes in the opposite direction uninjured, and in some instances forming long ridges of detritus along the surface. These are curious considerations, and it has become a question of much interest, by what means, and under what circumstances, such a current was produced. But in the present state of our knowledge, all that can be legitimately inferred from the diluvium is, that many portions of the northern regions of Europe and America were then under the sea, and that a strong current set over them.

Connected with the Diluvium is the history of *Ossiferous Caverns*, of which specimens singly exist at Kirkdale in Yorkshire, Gailenreuth in Franconia, and other places They occur in the calcareous strata, as the great caverns generally do, but have in all instances been naturally closed up till the recent period of their discovery. The floors are covered with what appears to be a bed of the diluvial clay, over which rests a crust of stalagmite, the result of the droppings from the roof since the time when the clay bed was laid down. In the instances above specified, and several others, there have been found, under the clay bed, assemblages of the bones of animals, of many various kinds. At Kirkdale, for example, the remains of twenty-four species were ascertained—namely, pigeon, lark, raven, duck, and partridge; mouse, water-rat, rabbit, hare, hippopotamus, rhinoceros, elephant, weasel, fox, wolf, deer, (three species,) ox, horse, bear, tiger, hyena. From many of the bones of the gentler of these animals being found in a broken state, it is supposed that the cave was a haunt of hyenas and other predaceous creatures, by which the smaller ones were here consumed. This must have been at a time antecedent to the floodings which produced the diluvium, since the bones are covered by a bed of that formation. It is impossible not to see here a very natural series of incidents. First, the cave is frequented by wild beasts, who make it a kind of charnel-house. Then, submerged in the current which has been spoken of, it receives a clay flooring from the waters containing that matter in suspension. Finally, raised from the water, but with no mouth to the open air, it remains unintruded on for a long series of ages, during which the clay flooring receives a new calcareous covering, from the droppings of the roof.

Our attention is next drawn to the erratic blocks or boulders, which in many parts of the earth are thickly strewn over the surface, particularly in the north of Europe. Some of these blocks are many tons in weight, yet are clearly ascertained to have belonged originally to situations at a great distance. Fragments, for example, of the granite of Shap Fell are found in every direction around to the distance of fifty miles, one piece being placed high upon Criffel Mountain, on the opposite side of the Solway estuary; so also are fragments of the Alps found far up the slopes of the Jura. There are even blocks on the east coast of England, supposed to have travelled from Norway. The only rational conjecture which can be formed as to the transport of such masses from so great a distance, is one which presumes them to have been carried and dropped by icebergs, while seas existed upon the space between their original and final sites. Icebergs do even now carry off such masses from the polar coasts, which, falling when the retaining ice melts, must take up situations at the bottom of the sea similar to those in which we find the erratic blocks of the present dry land.

While the diluvium and erratic blocks clearly suppose a part at least of the present land to have at one time been sunk to a considerable depth in the sea, there is another set of appearances which as manifestly show the steps by which the land was made afterwards to re-emerge from that element. These consist of *terraces*, which have been detected near, and at some distance inland from, the coast lines of Scandinavia, Britain, America, and other regions; being evidently ancient beaches, or platforms, on which the margin of the sea at one time rested. They have been observed at different heights above the present sea-level, from twenty to above twelve hundred feet; and in many places they are seen rising above each other in succession, to the number of three, four, and even more. The smooth flatness of these terraces, with generally a slight inclination towards the sea, the sandy composition of many of them, and, in some instances, the preservation of marine shells in the ground, identify them perfectly with existing sea-beaches, notwithstanding the cuts and scoopings which have at frequent intervals been effected in them by water-courses. The irresistible inference from the phenomena is, that the highest was first the coast line; then an elevation took place, and the second highest became so, the first being now raised into the air and thrown inland. Then, upon another elevation, the sea began to form, at its new point of contact with the land, the third highest beach, and so on down to the platform nearest to the present sea-beach. Phenomena of this kind become comparatively familiar to us, when we hear of evidence that the last sixty feet of the elevation of Sweden, and the last eighty-five of that of Chili, have taken place since man first dwelt in those countries; nay, that the elevation of the former country goes on at this time at the rate of about forty-five inches in a century, and that a thousand miles of the Chilian coast rose four feet in one night, under the influence of a powerful earthquake, so lately as 1822. Subterranean forces, of the kind then exemplified in Chili, supply a ready explanation of the whole phenomena, though some other operating causes have been suggested. In an inquiry on this point, it becomes of consequence to learn some particulars respecting the levels. Taking a particular beach, it is generally observed that the level continues the same along a considerable number of miles, and nothing like breaks or hitches has as yet been detected in any case. A second and a third beach are also observed to be exactly parallel to the first. These facts would seem to indicate quiet elevating movements, uniform over a large tract. It must, however, be remarked that the raised beaches at one part of a coast rarely coincide with those at another part forty or fifty miles off. We might suppose this to indicate a limit in that extent of the uniformity of the elevating cause; but it would be rash to conclude positively that such is the case. In the present sea, as is well known, there are different levels at different places, owing to the operation of peculiar local causes, as currents, evaporation, and the influx of large rivers into narrow-mouthed estuaries. The differences of level in the ancient beaches might be occasioned by some such causes. But, whatever doubt may rest on this minor point, enough has been ascertained to settle the main one, that we have in these platforms indubitable monuments of an elevation of the land from the sea, and the concluding great event of the geological history.

The idea of such a deep immersion of the land unavoidably suggests some considerations as to the effect which it might have upon terrestrial animal life. Some, regarding it as a complete submersion, argue that such life would be, on such an occasion, extensively, if not universally, destroyed. Nor was the idea of its universal destruction the less plausible, when it was believed that the present land animals are an entirely new set of species, introduced since the conclusion of the Tertiary Formation. It must now be owned that there are great objections, if not positive proof, against such an hypothesis. First, it is not true that the species of the tertiary epoch are all of them extinct. There are several—for example, a badger of the Miocene—which are not in the slightest degree distinguished from living species. Second, the specific distinctions alleged in a great number of cases between tertiary and existing animals are extremely slight, and such as we have no fixed principle by which to be assured, that they mark new species. Finally, we find the tertiary animals of America indicating an approximation to the characters of existing animals in that region, and tertiary animals of the other great continent equally approximating to those at present occupying it; showing that the demarcations of the present great zoological provinces had been already marked out, and have never been obliterated. There is therefore enough to justify us in believing that no entire submergence of the earth took place at the time of the Diluvium, though how nearly it might approach completeness we cannot say.

There are some other superficial formations, of less consequence on the present occasion than the diluvium—namely, lacustrine deposits, or filled-up lakes; alluvium, or the deposits of rivers beside their margins; deltas, the deposits made by great ones at their efflux into the sea; peat mosses; and the vegetable soil. The animal remains found in these generally testify to a zoology on the verge of that now prevailing, or melting into it, there being included many species which still exist. In a lacustrine deposit at Market-Weighton, in the Vale of York, there have been found bones of the elephant, rhinoceros, bison, wolf, horse, felis, deer, birds, all or nearly all belonging to extinct species; associated with thirteen species of land and fresh-water shells, "exactly identical with types now living in the vicinity." In similar deposits in North America, are remains of the mammoth, mastodon, buffalo, and other animals of extinct and living types. In short, these superficial deposits show precisely such remains as might be expected from a time at which the present system of things (to use a vague but not unexpressive phrase) obtained, but yet so far remote in chronology as to allow of the dropping of many species, through familiar causes, in the interval. Still, however, there is no authentic or satisfactory instance of human remains being found, except in deposits obviously of very modern date; a tolerably strong proof that the creation of our own species is a comparatively recent event, and one posterior (generally speaking) to all the great natural transactions which have been here described.

GENERAL CONSIDERATIONS

RESPECTING

THE ORIGIN OF THE ANIMATED TRIBES.

THUS concludes the wondrous chapter of the earth's history which is told by geology. It takes up our globe at the period when its original incandescent state had nearly ceased; conducts it through what we have every reason to believe were vast, or at least very considerable, spaces of time, in the course of which many superficial changes took place, and vegetable and animal life was gradually devolved; and drops it just at the point when man was apparently about to enter on the scene. The compilation of such a history, from materials of so extraordinary a character, and the powerful nature of the evidence which these materials afford, are calculated to excite our admiration, and the result must be allowed to exalt the dignity of science, as a product of man's industry and his reason.

If there is anything more than another impressed on our minds by the course of the geological history, it is, that the same laws and conditions of nature now apparent to us have existed throughout the whole time, though the operation of some of these laws may now be less conspicuous than in the early ages, from some of the conditions having come to a settlement and a close. That seas have flowed and ebbed, and winds disturbed their surfaces, in the time of the secondary rocks, we have proof on the yet preserved surfaces of the sands which constituted margins of the seas in those days. Even the fall of wind-slanted rain is evidenced on the same tablets. The washing down of detached matter from elevated grounds, which we see rivers constantly engaged in at the present time, and which is daily shallowing the seas adjacent to their mouths, only proceeded on a greater scale in earlier epochs. The volcanic subterranean force, which we see belching forth lavas on the sides of mountains, and throwing up new elevations by land and sea, was only more powerfully operative in distant ages. To turn to organic nature, vegetation proceeded then exactly as now.

The very alternation of the seasons has been read in unmistakable characters in sections of the trees of those days, precisely as it might be read in a section of a tree cut down yesterday. The system of prey amongst animals flourished throughout the whole of the pre-human period; and the adaptation of all plants and animals to their respective spheres of existence was as perfect in those early ages as it is still.

But, as has been observed, the natural laws, though essentially unchanging, may operate with greater force at one time than at another, and may be modified by attendant conditions. In the carbonigenous era, dry land seems to have consisted only of clusters of islands, and the temperature was much above what now prevails at the same places. Volcanic forces, and perhaps also the disintegrating power, have been on the decrease since the first, or we have at least long enjoyed an exemption from such paroxysms of the former, as for certain prevailed at the close of the coal formation in England and throughout the tertiary era. The surface has also undergone a gradual progress by which it has become always more and more variegated, and thereby fitted for the residence of a higher class of animals.

In pursuing the progress of the development of both plants and animals upon the globe, we have seen an advance in both cases, from simple to higher forms of organization. In the botanical department, we have first sea, afterwards land plants; and amongst these, the simpler (cellular and cryptogamic) before the more complex. In the department of zoology, we see, first, traces all but certain of infusoria; then polypiaria, crinoidea, and some of the articulata and mollusca; afterwards higher forms of the most of these humble classes of animals; and it appears that such existed for ages before there were any higher types of being. The first step forward gives fishes, the humblest class of the vertebrata; and these are of the cartilaginous orders, and marked by several other traits of inferior organization.

Afterwards come land animals, of which the first are reptiles, universally allowed to be the type next in advance from fishes, and to be connected with these by the links of an insensible gradation. From reptiles we advance to birds, and thence to mammalia, the first of which are marsupialia, acknowledgedly low forms in their class. That there is thus a progress of some kind, the most superficial glance at the geological history is sufficient to convince us. Indeed, the doctrine of the gradation of animal forms has received a remarkable support from the discoveries of this science, as several types formerly wanting to a completion of the series have been found in a fossil state.*

Fossil history has no doubt some obscure and difficult passages, but they are not more than might be expected, when we consider certain obvious circumstances attending, first, the inhumation, and afterwards the discovery or investigation of fossils. These are unquestionably of a nature to forbid completeness and regularity in the details of the fossil history, though not to affect the great leading facts. For example, some animal and vegetable forms are of too slight a structure to admit of their being preserved in rocks, especially those of an early kind, which have almost all been greatly altered by exposure to heat. It is also certain that in the primeval, as in the present seas, various animals affected various localities, according to peculiarities of depth, temperature, and opportunities of feeding; so that no particular group found as fossils in one geogra-

* Intervals in the series were numerous in the department of the pachydermata; many of these gaps are now filled up from the extinct genera found in the tertiary formation.

phical space can be received as a perfect sample of the whole animal kingdom as then existing. We are also, it must be observed, in the infancy of geological research; every year is bringing us new light respecting the details of fossil history from various regions of the globe. Then, we are but doubtfully informed on the affinities and comparative organization of animals: beyond the general facts with regard to sub-kingdoms and classes, zoological science does not at present speak conclusively on the subject of gradation. All these things being considered, we are not entitled, upon any blanks or confusions which may at present appear in these details, to ground decisive objections to that idea of progress in animated nature which palæontology in the main so plainly teaches. Thus, for the facts—if they be such—that we have at first no traces of plants simpler than fuci, or animals below the polypiaria and crinoidea, it may be said in explanation that such organic forms, supposing they had existed in the greatest profusion, had hardly any chance of leaving memorials of their substance in the metamorphic rocks. When we find, in the English protozoic rocks, highly organized carnivorous mollusks side by side with lower orders of invertebrata, we may point to America, where the protozoic rocks present no such exalted forms, and ask if the very earliest fossiliferous rocks, the true protozoic formation, have as yet been anywhere found. Finally, with respect to a reptilian dentition appearing at an early period among fishes,—lacertine reptiles with no apparent affinity to these being the first fossils as yet discovered in that class,—footsteps of wading and running birds coming before any trace of lower ornithic forms, and so forth,—I may refer to the partiality of all geological discovery, and the doubtfulness of all existing classification. In a succeeding chapter, such a view of the animal kingdom upon a strictly natural arrangement will be presented, as it is hoped will tend to abate many of these difficulties. For the rest, we may hope that more ample geological research will in time afford a sufficient explanation. In the mean time, the great facts stand free of all doubt. Invertebrate are inferior to vertebrate animals; the vertebrate are ranked in this order—fishes, reptiles, birds, mammals: such, too, is the order in which these creatures occur in the chronology of rock formations. So also it is undoubted that the earliest plants are acotyledons, the next monocotyledons, the next dicotyledons, and such is the acknowledged order of comparative organization in that kingdom. It is equally important to remark that, amidst all the imperfection of the ancient series of animal forms, arising first from partiality in preservation, and second from partiality in discovery, many of the details as to families harmonize with even the obscure ideas of classification which at present reign amongst naturalists. Hence, it appears to me, there is strong reason to apprehend that, in the history of animated nature, a progress of some kind has been observed.

Now this progress involves two considerations: it is a progress which has observed a rule of time; it is a progress which has observed a rule of advancing organization. Here we have two remarkable and perfectly definite ideas respecting the history of animated nature. A third rule will afterwards be seen to have affected organization—namely, external conditions in the earth itself; but this we are not at present called upon to consider. Fixing attention meanwhile upon the facts that the organic kingdoms have been produced in the course of a long space of time and by steps of progressive improvement or advance, we readily see that, while the question of the Divine authorship of the universe remains exactly as it was, we can no longer rest satisfied with the common notions regarding the mode of working observed by that power in this instance. We call in question, not merely the simple idea of the unenlightened mind, that God fashioned all in the manner of an artificer seeking by special means to produce special effects, but even the doctrine in vogue amongst men of science, that "creative fiats" were required for each new class, order, family, and species of organic beings, as they successively took their places upon the globe, or as the globe became gradually fitted for their reception. For, if such fiats were the mode of the Deity's operation, how should there have been oceans existing for unreckonable ages without fish, or dry land without land animals? How should the dry land have afterwards been possessed for ages by reptiles, without any superior animals, notwithstanding that we now find mammifers capable of living wherever reptiles can exist? Or why should there finally have been an age of inferior mammifers, without the presence of the highest order of being? And, in such a case, why should the supposed fiats have evoked being in this order, considering that there was nothing we can see in the condition of the earth in early ages to prevent fish from living in the sea before invertebrates, or mammifers from flourishing in the carboniferous woods before the rise of reptiles? It is startling, too, to think of this miraculous procedure being introduced in the midst of a system of things which was in every respect strictly natural—while seas were wearing down cliffs and forming new strata in their bosom, while forests grew and decayed, while the wind blew and the rain fell, exactly as at the present day. Nor is it less startling to regard it as following upon a series of operations much grander both in their character and results, which we know to have proceeded in the manner of natural law. The eternal Sovereign arranges a planetary or an astral system by dispositions imparted primordially to matter; but he has to give a particular heed to the formation of the few corals and shell-fish in the Cambrian seas; he has by a new fiat to add fish, afterwards reptiles—birds—mammifers; and not only these great classes, but each particular species of which those classes are composed! In short, we are called upon by this theory to believe in a system of discrepancies and contradictions, to which it seems impossible for any candid and awakened mind to give credence.

It is under a humble sense of the difficulties which beset the presumption of extraordinary or extra-natural causes for the origin of the organic kingdoms; it is, I may say, in obedience to the hints and beckonings of nature herself: that I am led to inquire if there be any insurmountable obstacle to our embracing a rational modification of the opposite theory—namely, that species were introduced upon our globe by virtue of primordial arrangements having that object in view, and in which the Deity was only present and active in the sense in which he is so in all the phenomena of nature.

The hypothesis of an organic creation by natural law has, within the last two thousand years, been presented in various forms to mankind, but never has met with a hearty reception, the unenlightened mind greatly preferring the idea of a special mode of action suited to every contingency, and the philosophical mind resting generally at some point in this preference, to which it reconciled itself by reflecting that our knowledge did not enable us to arrive at a satisfactory conclusion with regard to the mode of the procedure implied. One great cause for the opposition made to the hypothesis was its being usually brought forward as something superseding the idea of an intelligent creative power, or at least being supposed to have this effect, so that there seemed only a choice between the doctrine of a deity acting for every occasion, and that of a cold and remote abstraction called nature. In reality, the hypothesis does not necessarily, in the least degree, supersede the Deity as the author and ruler of his works. It merely argues for one particular mode of acting instead of another, in one of the departments of his creation. It only would take one more section of the mundane economy out of the region of the supernatural in which ignorance would retain it all, and place it within that field of order to which it has been the tendency of all science to transfer the phenomena of the world.

As to any presumed bearing of the theory upon the character of the Creator, it might be enough to say that whatever it does in this respect, is certainly done in an equal degree by every other fact of nature which scientific knowledge has induced us to consider upon the footing of law. But I am unwilling to rest content with such a defence, and will add, that every advance we make in this direction, when rightly considered, affects our ideas of deity for the better. It is nothing less than a mean view of the Great Author, to suppose him obliged to come in on frequent occasions with new fiats or special interferences. It detracts immensely from his foresight, the grandest of all his attributes. The opposite doctrine supposes a much higher kind of power, exerted in an infinitely more sublime manner; not, be it always observed, a power which gave a first impulse and then retired into inaction, but one which, having pre-determined the universal scheme, sustains the whole of its complicated operations, in serene, immutable energy, present in the simplest phenomena as in the highest, never for a moment absent or asleep. "If," says Dr. Buckland, contemplating the possible establishment of this doctrine—"if the properties adopted by the elements at the moment of their creation adapted them beforehand to the infinity of complicated useful purposes which they have already answered, and may have still further to answer, under many dispensations of the material world, such an aboriginal constitution, *so far from superseding an intelligent agent, would only exalt our conceptions of the consummate skill and power that could comprehend such an infinity of future uses under future systems, in the original groundwork of his creation."*

It may, however, be objected that the ideas of Christian nations on this subject are derived from an authority which does not admit of human reasonings upon their tendency or bearing. And, certainly, were that authority more forcible in its enunciations on this point than it is with regard to many other philosophical doctrines now fully admitted—such as the solar system of Copernicus—we might well feel that there was a difficulty in our path. But in reality such is not the case. There is hardly one of the admitted natural laws against which it would not be possible to bring, equally as against this, a powerful show of scriptural evidence, if we were to adopt in all those cases similar principles of interpretation. Yet more; if we look carefully, and with an awakened mind, into the scriptural record, we shall find the creative procedure represented in the first place, not as consisting in special efforts of the deity, but as resulting from commands or expressions of will. Let there be light—let there be a firmament—let the dry land appear—let the earth bring forth grass, the herb, the tree—let the sea bring forth the moving creature that hath life—let the earth bring forth the living creature after his kind—these are the terms in which the principal acts are described, conforming exactly to our ideas of natural laws. The additional expressions, God made the firmament—God made the beast of the earth, &c.—occur subordinately, and in comparatively few instances; they do not necessarily convey a different idea of the mode of creation, and indeed only appear as alternative phrases in the usual duplicative style of the east. There is thus, in what most persons will probably acknowledge as the source of their ideas on this subject, certainly nothing like the decided affirmation of the doctrine of special exercise which has been assumed. Upon what, then, does this doctrine depend for the superior respectability arrogated for it, beyond its having been the first and the long entertained presumption of man—the presumption of his infancy and non-age? Was not the doctrine of the centrality of the earth exactly such a presumption till two centuries ago? equally a natural first impression of the human mind, and equally assumed to be sanctioned by Scripture, and yet proved unsound at last and now wholly given up?

It will be the object of some of the subsequent chapters to bring the special facts of several of the sciences to the further illustration and support of this doctrine. Meanwhile, we may remark what is perhaps the most powerful argument for it, of a general kind—namely, its harmony with the presumable associated phenomena. First, it agrees with the idea of planet creation by law. Secondly, upon this supposition, all that geology tells us of the succession of species becomes natural and intelligible. We see occasion for Time to evolve the vegetable and animal kingdoms, and we no longer can wonder that the details of these kingdoms came in the order of progressive organization. Those changes, also, of species which occur in the geological record,—so extensive at some points as to suggest the idea of a complete renewal of organic life over the globe,—become explicable as for the most part mere transmutations of one into another, effected by the operation of natural causes, whether those which cause an advance in grade, or those which we shall trace as productive of only external modifications.

Nor are we only to account for the origination

of organic being upon this little planet, third of a series which is but one of hundreds of thousands of series, the whole of which again form but one portion of an apparently infinite globe-peopled space, where all seems analogous. We have to suppose, that every one of these numberless globes is either a theatre of organic being, or in the way of becoming so. This is a conclusion which every addition to our knowledge makes only the more irresistible. Is it conceivable, as a fitting mode of exercise for creative intelligence, that it should be constantly paying a special attention to the creation of species, as they may be required in each situation throughout those worlds, at particular times? Is such an idea accordant with our general conception of the dignity, not to speak of the power, of the Great Author? Yet such is the notion which we must form, if we adhere to the doctrine of special exercise. Let us see, on the other hand, how the doctrine of a creation by law agrees with this expanded view of the organic world.

Unprepared as most men may be for such an announcement, there can be no doubt that we are able, in this limited sphere, to form some satisfactory conclusions as to the plants and animals of those other spheres which move at such immense distances from us. Suppose that the first persons of an early nation who made a ship and ventured to sea in it, observed, as they sailed along, a set of objects which they had never before seen—namely, a fleet of other ships—would they not have been justified in supposing that those ships were occupied, like their own, by human beings possessing hands to row and steer, eyes to watch the signs of the weather, intelligence to guide them from one place to another—in short, beings in all respects like themselves, or only showing such differences as they knew to be producible by difference of climate and habits of life? Precisely in this manner we can speculate on the inhabitants of remote spheres. We see that matter has originally been diffused in one mass, of which the spheres are portions. Consequently, inorganic matter must be presumed to be everywhere the same, although probably with differences in the proportions of ingredients in different globes, and also some difference of conditions. Out of a certain number of the elements of inorganic matter are composed organic bodies, both vegetable and animal; such must be the rule in Jupiter and in Sirius, as it is here. We, therefore, are all but certain that herbaceous and ligneous fibre, that flesh and blood, are the constituents of the organic beings of all those spheres which are as yet seats of life. Gravitation we see to be an all-pervading principle: therefore there must be a relation between the spheres and their respective organic occupants, by virtue of which they are fixed, as far as necessary, on the surface. Such a relation, of course, involves details as to the density and elasticity of structure, as well as size of the organic tenants, in proportion to the gravity of the respective planets —peculiarities, however, which may quite well consist with the idea of a universality of certain types, such as we see exemplified upon earth. We come to comparatively matter of detail, when we advert to heat and light; yet it is important to consider that these are universal agents, and that, as they bear marked relations to organic life and structure on earth, they may be presumed to do so in other spheres also. The considerations as to light are particularly interesting, for, on our globe, the structure of one important organ, almost universally distributed in the animal kingdom, is in direct and precise relation to it. Where there is light there will be eyes, and these, in other spheres, will be the same in all respects as the eyes of tellurian animals, with only such differences as may be necessary to accord with minor peculiarities of condition and of situation. It is but a small stretch of the argument to suppose that, one conspicuous organ of a large portion of our animal kingdom being thus universal, a parity in all the other organs—species for species, class for class, kingdom for kingdom—is highly likely, and that thus the inhabitants of all the other globes of space bear not only a general, but a particular resemblance to those of our own.

Assuming that organic beings are thus universally distributed, the idea of their having all come into existence by the operation of laws everywhere applicable, is strictly conformable to the principle laid down for our own limited sphere. As one set of laws produced all orbs, their motions and geognostic arrangements, so one set of laws overspread them all with life. The whole productive or creative arrangements thus appear in perfect unity.

PARTICULAR CONSIDERATIONS

RESPECTING

THE ORIGIN OF THE ANIMATED TRIBES.

This concludes the general argument for organic creation by law, as preferable to that which, equally on hypothetic grounds, presumes a special exertion of divine power for each detached portion of that series of phenomena. We are now to inquire what organic nature herself says with respect to her origin; that is to say, whether she declares most loudly for the special or the general exercise of divine power in the production of her many tribes.

At the very first it may be frankly admitted, that there is no proof which has been satisfactory to the philosophical world, either of the origination of life from inorganic matter, or of the commencement of a new species, having ever once taken place since man first breathed on earth. It is not indeed strictly disproved that life is occasionally originated at the present time otherwise than by parentage; but neither has it been proved to the satisfaction of men of science that any such phenomenon ever did take place. There is here an appearance of strong presumption against the hypothesis of organic creation by law, and yet why should it be so considered? The great work of the peopling of this globe with living species is mainly a fact accomplished: the highest known species came as a crowning effect thousands of years ago. The work being thus, to all appearance, finished, we are not necessarily to expect that the origination of life and of species should be conspicuously exemplified in the present day. We are rather to expect that the vital phenomena presented to our eyes should mainly, if not entirely, be limited to a regular and unvarying succession of races by the ordinary means of generation. This, however, is no more an argument against a time when phenomena of the first kind prevailed, than it would be a proof against the fact of a mature man having once been a growing youth, that he is now seen growing no longer. We might consider the primitive production of species either as one phenomenon of the nature of the development of an individual embryo, and that phenomenon as past, just as the individual creation is perfected at birth, or as expressly and wholly a consequence of conditions, which being temporary the results were temporary also. Perhaps, from the occupation of all the great geographical provinces with a full suite of the forms of life, a new development may have hardly any chance of being now drawn forth, and none of being advanced to any extent, even though the same life-creating laws be still in operation. Or the operations of these laws might be observant of times, and these of rare occurrence, so that hundreds of human generations may pass without an opportunity of witnessing their effects. However it may actually have been, assuredly the most rigid disproof of what is called spontaneous generation as a fact of our day, could be no conclusive argument against a law creation of organisms which is announced as having taken place many thousand years ago.

In the existing circumstances—the organic creation being mainly completed—the whole phenomenon having passed, unnoted by any intellect that could chronicle its mode—it cannot be expected that we should be able to present anything beyond mere vestiges or faint memorials of the procedure of God in this instance. But certainly, if several of the natural sciences afford us such vestiges, and none of them make positive affirmation to the contrary purpose, a considerable support must be given to the hypothesis shown on general grounds to be most likely—namely, that God employed the mandates of his unvarying will in communicating the breath of life, as well as in the forming of globes and arranging them in space.

Crystallization is confessedly a phenomenon of inorganic matter; yet the simplest rustic observer is struck by the resemblance which the examples of it left upon a window by frost bear to vegetable forms. In some crystallizations the mimicry is beautiful and complete; for example, in the well-known one called the *Arbor Dianæ.* An amalgam of four parts of silver and two of mercury being dissolved in nitric acid, and water equal to thirty weights of the metals being added, a small piece of soft amalgam of silver, suspended in the solution, quickly gathers to itself the particles of the silver of the amalgam, which form upon it a *crystallization precisely resembling a shrub.* Vegetable figures are also presented in some of the most ordinary appearances of the electric fluid. In the marks caused by positive electricity, or which it leaves in its passage, we see the ramifications of a tree, as well as of its individual leaves; those of the negative, recal the bulbous or the spreading root, according as they are clumped or divergent. These phenomena seem to indicate that the electric energies have had something to do in determining the forms of plants. That they are intimately connected with vegetable life is indubitable, for germination will not proceed in water charged with negative electricity, while water charged positively greatly favours it; and a garden sensibly increases in luxuriance, when a number of conducting rods are made to terminate in branches over its beds. With regard to the resemblance of the ramifications of the branches and leaves of plants to the traces of the positive electricity, and that of the roots to the negative, it is a circumstance calling for especial remark, that the atmosphere, particularly its lower strata, is generally charged positively, while the earth is always charged negatively. The correspondence here is curious. A plant thus appears as a thing formed on the basis of a natural electrical operation — *the brush* realized. We can thus suppose the various forms of plants as, immediately, the result of a law in electricity, variously affecting them according to their organic character, or respective germinal constituents. In the poplar, the brush is unusually vertical, and little divergent; the reverse in the beech: in the palm, a pencil has proceeded straight up for a certain distance, radiates there, and turns outwards and downwards; and so on. We can here see at least traces of secondary means by which the Almighty Deviser might establish all the vegetable forms with which the earth is overspread.*

Vegetable and animal bodies are mainly composed of the same four simple substances or elements—carbon, oxygen, hydrogen, and nitrogen. The first combinations of these in animals are into what are called proximate principles, as albumen, fibrin, &c., out of which the structure of the animal body is composed. Now it is acknowledged by Dr. Daubeny, that in the combinations forming the proximate principles there is no chemical peculiarity. "It is now certain," he says, "that the same simple laws of composition pervade the whole creation; and that, if the

* See Appendix, E.

organic chemist only takes the requisite precautions to avoid resolving into their ultimate elements the proximate principles upon which he operates, the result of his analysis will show that they are combined precisely according to the same plan as the elements of mineral bodies are known to be."* A particular fact is here worthy of attention. "The conversion of fecula into sugar, as one of the ordinary processes of vegetable economy, is effected by the production of a secretion termed *diastase*, which occasions both the rupture of the starch vesicles, and the change of their contained gum into sugar. This diastase may be separately obtained by the chemist, and it acts as effectually in his laboratory as in the vegetable organization. He can also imitate its effects by other chemical agents."† The writer quoted below adds, "No reasonable ground has yet been adduced for supposing that, if we had the power of bringing together the elements of any organic compound, in their requisite states and proportions, the result would be any other than that which is found in the living body."

It is much to know the elements out of which

* Supplement to the Atomic Theory.

† Carpenter on Life; Todd's Cyclopædia of Physiology.

organic bodies are composed. It is something more to know their first combinations, and that these are simply chemical. How these combinations are associated in the structure of living bodies is the next inquiry, but it is one to which as yet no satisfactory answer can be given. The investigation of the minutiæ of organic structure by the microscope is of such recent origin, that its results cannot be expected to be very clear. Some facts, however, are worthy of attention with regard to the present inquiry. It is ascertained that the basis of all vegetable and animal substances consists of nucleated cells; that is, cells having granules within them. Nutriment is converted into these before being assimilated by the system. The tissues are formed from them. The ovum destined to become a new creature, is originally only a cell with a contained granule. We see it acting this reproductive part in the simplest manner in the cryptogamic plants. "The parent cell, arrived at maturity by the exercise of its organic functions, bursts, and liberates its contained granules. These, at once thrown upon their own resources, and entirely dependent for their nutrition on the surrounding elements, develop themselves into new cells, which repeat the life of their original. Amongst the higher tribes of the cryptogamia, the reproductive cell does not burst, but the first cells of the new structure are developed within it, and these gradually extend, by a similar process of multiplication, into that primary leaf-like expansion which is the first formed structure in all plants."* *Here the little cell becomes directly a plant, the full formed living being.* It is also worthy of remark that, in the sponges, (an animal form,) a gemmule detached from the body of the parent, and trusting for sustentation only to the fluid into which it has been cast, becomes, without further process, the new creature. Further, it has been recently discovered by means of the microscope, that there is, as far as can be judged, a perfect resemblance between the ovum of the mammal tribes, during that early stage when it is passing through the oviduct, and the young of the infusory animalcules. One of the most remarkable of these, the *volvox globator*, can hardly be distinguished from the germ which, after passing through a long fœtal progress, becomes a complete mammifer, an animal of the highest class. It has even been found that both are alike provided with those *cilia*,

* Carpenter's Report on the results obtained by the Microscope in the study of Anatomy and Physiology, 1843.

which, producing an appearance of revolving motion, is partly the cause of the name given to this animalcule. These resemblances are the more entitled to notice, that they were made by various observers, distant from each other at the time.* It has likewise been noted that the globules of the blood are reproduced by the expansion of contained granules; they are, in short, *distinct organisms multiplied by the same fissiparous generation.* So that all animated nature may be said to be based on this mode of origin; *the fundamental form of organic being is a cell, having new cells forming within itself,* by which it is in time discharged, and which are again followed by others and others, in endless succession. It is of course obvious that, if these cells could be produced by any process from inorganic elements, we should be entitled to say that the fact of a transit from the inorganic into the organic had been witnessed in that instance; the possibility of the commencement of animated creation by the ordinary laws of nature might be considered as established. Now it was announced some years ago by Prevost and Dumas, that *globules could be produced in albumen*

* See Dr. Martin Barry on Fissiparous Generation; Jameson's Journal, Oct. 1843.

by electricity. If, therefore, these globules be identical with the cells which are now held to be reproductive, it might be said that the production of albumen by artificial means is the only step in the process wanting. This has not yet been effected; but it is known to be only a chemical process, the mode of which may be any day discovered in the laboratory.*

There is here rather a looking forward to something which science may yet define and realize to us, than the statement of any facts positively evidencing the subjection of what we call life to common physical laws. It is, however, indubitable that the tendency of science has long been towards the abandonment of even the modified idea of a "vital principle" as a distinct natural force, and to the conclusion that living structures result from the action of a multitude of such forces in combination—"gravity, cohesion, elasticity, the agency of the imponderables, and all other powers which operate both on masses and atoms." Professor Draper, of New York, in making this statement, says—"It is astonishing that in our days the ancient system which excludes all connexion with natural philosophy and chemistry, and depends

* See Appendix, F.

on the fictitious aid of a visionary force, should continue to exist; a system which at the outset ought to have broken down by the most common considerations, such as those connected with the mechanical principles involved in the bony skeleton, the optical principles in the construction of the eye, or the hydraulic action of the valves of the heart."*

With respect to the doctrine of spontaneous generation, I may repeat that a decided negative put to it by science could not be in any degree conclusive against our hypothesis, seeing that the phenomenon might be presumed to be one of time, and that the time might be past, or of conditions, and the conditions may have ceased. Nevertheless, as we see such laws as those which produced degradation and upheavals in early times still acting with a faint representation of their former activity, it seems not unlikely that we should still see some remnants, or partial and occasional workings of the life-creating energy amidst a system of things generally stable and at rest. Are there, then, any such remnants to be traced in our own day, or during man's existence upon earth?

* Treatise on the Forces which produce the Organization of Plants. New York, 1844.

If there be, it clearly would form a strong evidence in favour of the doctrine of organic creation by law, as what now takes place upon a confined scale, and in a comparatively casual manner, may have formerly taken place on a great scale, and as the proper and eternity-destined means of supplying a vacant globe with suitable tenants.

Most scientific men of approved reputation would certainly answer in the negative. The professed reason is that, in a great number of instances where the superficial observers of former times assumed a non-generative origin for life, (as in the celebrated case in Virgil's fourth Georgic,) either the direct contrary has been ascertained, or exhaustive experiments have left no alternative from the conclusion that ordinary generation did take place, albeit in a manner which escapes observation. Finding that an erroneous assumption has been formed in many cases, modern inquirers have not hesitated to assume that there can be no case in which generation is not concerned. Now their conclusion may be right, but it clearly is not one beyond question; and it is equally true that the explanations suggested in difficult cases are often far from being satisfactory. There are several persons eminent in science who profess at least to find great difficulties in accepting the doctrine of invariable generation. One * has stated several considerations arising from analogical reasoning, which appear to him to throw the balance of evidence in favour of the primitive production of infusoria, the vegetation called mould, and the like. One seems to be of great force; namely, that the animalcules, which are supposed (altogether hypothetically) to be produced by ova, are afterwards found increasing their numbers, not by that mode at all, but by division of their bodies. If it be the nature of these creatures to propagate in this splitting or fissiparous manner, how could they be communicated to a vegetable infusion? It has been shown by the opponents of this theory, that when a vegetable infusion is debarred from the contact of the atmosphere, by being closely sealed up or covered with a layer of oil, no animalcules are produced; but it has been said, on the other hand, that the exclusion of the air may prevent some simple condition necessary for the aboriginal development of life—and this cannot be denied. Perhaps the prevailing doctrine is in nothing placed in greater diffi-

* Dr. Allen Thomson, in the article *Generation*, in Todd's Cyclopædia of Anatomy and Physiology.

culties than it is with regard to the entozoa, or creatures which live within the bodies of others. These creatures do, and apparently can, live nowhere else than in the interior of other living bodies, where they generally take up their abode in the viscera, but also sometimes in the chambers of the eye, the interior of the brain, the serous sacs, and other places having no communication from without. Some are viviparous, others oviparous. Of the latter it cannot reasonably be supposed that the ova ever pass through the medium of the air, or through the blood-vessels, for they are too heavy for the one transit, and too large for the other. Of the former, it cannot be conceived how they pass into young animals—certainly not by communication from the parent, for it has often been found that entozoa do not appear in certain generations of a human family, and some of peculiar and noted character have only appeared at rare intervals, and in very extraordinary circumstances. A candid view of the less popular doctrine, as to the origin of this humble form of life, is taken by a distinguished living naturalist. "To explain the beginning of these worms within the human body, on the common doctrine that all created beings proceed from their likes, or a primordial egg, is so difficult, that the moderns have been driven to speculate, as our fathers did, on their spontaneous birth; but they have received the hypothesis with some modification. Thus it is not from putrefaction or fermentation that the entozoa are born, for both of these processes are rather fatal to their existence, but from the aggregation and fit apposition of matter which is already organized, or has been thrown from organized surfaces. * * Their origin in this manner is not more wonderful or more inexplicable than that of many of the inferior animals from sections of themselves. * * Particles of matter fitted by digestion, and their transmission through a living body, for immediate assimilation with it, or flakes of lymph detached from surfaces already organized, seem neither to exceed nor fall below that simplicity of structure which favours this wonderful development; and the supposition that, like morsels of a planaria, they may also, when retained in contact with living parts, and in other favourable circumstances, continue to live and be gradually changed into creatures of analogous conformation, is surely not so absurd as to be brought into comparison with the Metamorphoses of Ovid. * * We think the hypothesis is also supported in some degree by the fact, that the origin of the entozoa is favoured by all causes which tend to disturb the equality between the secerning and absorbent systems."* Here particles of organized matter are suggested as the germinal original of distinct and fully organized animals, many of which have a highly developed reproductive system. How near such particles must be to the inorganic form of matter may be judged from what has been said within the last few pages. If, then, this view of the production of entozoa be received, it must be held as in no small degree favourable to the general doctrine of an organic creation by law.†

There is another series of facts, akin to the above, and which deserve not less attention. The pig, in its domestic state, is subject to the attacks of a hydatid, from which the wild animal is free; hence the disease called measles in pork. The domestication of the pig is of course an event subsequent to the origin of man; indeed, comparatively speaking, a recent event. Whence, then, the first progenitor of this hydatid? So also there is a tinea which attacks dressed wool, but never touches it in its unwashed state. A par-

* Article "Zoophytes," Encyclopædia Britannica, 7th edition.

† See Appendix, G.

ticular insect disdains all food but chocolate, and the larva of the *oinopota cellaris* lives nowhere but in wine and beer, all of these being articles manufactured by man. There is likewise a fish called the *pymelodes cyclopum*, which is only found in subterranean cavities connected with certain specimens of the volcanic formation in South America, dating from a time posterior to the arrangements of the earth for our species. Whence the first pymelodes cyclopum? Will it, to a geologist, appear irrational to suppose that, just as the pterodactyle was added as a new offshoot from the animal stock, in the era of the new red sandstone, when the earth had become suited for such a creature, so may these creatures have been added when media suitable for their existence arose, and that such phenomena may take place any day, the only cause for their taking place seldom being the rarity of the rise of new physical conditions on a globe which seems to have already undergone the principal part of its destined mutations?

Between such isolated facts and the greater changes which attended various geological eras, it is not easy to see any difference, besides simply that of the scale on which the respective phenomena took place, as the throwing off of one copy from an engraved plate is exactly the same process as that by which a thousand are thrown off. To Creative Providence, we may well conceive, the numbers of such phenomena, the time when, and the circumstances under which they take place, are indifferent matters. The Eternal One has arranged for everything beforehand, by the operation of laws of his appointment, himself being ever present in all things. We can even conceive that man, in his many doings upon the surface of the earth, may occasionally, without his being aware of it, or otherwise, act as an instrument in preparing the association of conditions under which the creative laws work; and perhaps some instances of his having acted as such an instrument have actually occurred in our own time.

I allude, of course, to the experiments conducted a few years ago by Mr. Crosse, which seemed to result in the production of a heretofore unknown species of insect in considerable numbers. Various causes have prevented these experiments and their results from receiving candid treatment, but they may perhaps be yet found to have opened up a new and most interesting chapter of nature's mysteries. Mr. Crosse was pursuing some experiments in crystallization, causing a powerful voltaic battery to operate upon a saturated solution of silicate of potash, when the insects unexpectedly made their appearance. He afterwards tried nitrate of copper, which is a deadly poison, and from that fluid also did live insects emerge. Discouraged by the reception of his experiments, Mr. Crosse soon discontinued them; but they were some years after pursued by Mr. Weekes, of Sandwich, with precisely the same results. This gentleman, besides trying the first of the above substances, employed ferro-cyanate of potassium, on account of its containing a larger proportion of carbon, the principal element of organic bodies; and from this substance the insects were produced *in increased numbers.* A few weeks sufficed for this experiment, with the powerful battery of Mr. Crosse: but the first attempts of Mr. Weekes required about eleven months, a ground of presumption in itself that the electricity was chiefly concerned in the phenomenon. The changes undergone by the fluid operated upon, were in both cases remarkable, and nearly alike. In Mr. Weekes's apparatus, the silicate of potash became first turbid, then of a milky appearance;

round the negative wire of the battery, dipped into the fluid, there gathered a quantity of *gelatinous matter*. From this Mr. Weekes observed one of the insects in the very act of emerging, immediately after which it ascended to the surface of the fluid, and sought concealment in an obscure corner of the apparatus. The insects produced by both experimentalists seem to have been the same, a species of acarus, minute and semi-transparent, and furnished with long bristles, which can only be seen by the aid of the microscope. It is worthy of remark, that some of these insects, soon after their existence had commenced, were found to be likely to extend their species. They were sometimes observed to go back to the fluid to feed, and occasionally they devoured each other.*

The reception of novelties in science must ever be regulated very much by the amount of kindred or relative phenomena which the public mind already possesses and acknowledges, to which the new can be assimilated. A novelty, however true, if there be no received truths with which it

* See a Pamphlet circulated by Mr. Weekes, in 1842. For a detail of further and more conclusive experiments, reference may be made to *Explanations, forming a Supplement to the Vestiges, &c.*

can be shown in harmonious relation, has little chance of a favourable hearing. In fact, as has been often observed, there is a measure of incredulity from our ignorance as well as from our knowledge, and if the most distinguished philosopher three hundred years ago had ventured to develop any striking new fact which only could harmonize with the as yet unknown Copernican solar system, we cannot doubt that it would have been universally scoffed at in the scientific world, such as it then was, or, at the best, interpreted in a thousand wrong ways in conformity with ideas already familiar. The experiments above described, finding a public mind which had never discovered a fact or conceived an idea at all analogous, were of course ungraciously received. It was held to be impious even to surmise that animals could have been formed through any instrumentality of an apparatus devised by human skill. The more likely account of the phenomena was said to be, that the insects were only developed from ova, resting either in the fluid, or in the wooden frame on which the experiments took place. On these objections the following remarks may be made. The supposition of impiety arises from an entire misconception of what is implied by an aboriginal creation of insects. The experimentalist could never be considered as the author of the existence of these creatures, except by the most unreasoning ignorance. The utmost that can be claimed for, or imputed to him is, that he arranged the natural conditions under which the true creative energy—that flowing from the primordial appointment of the Divine Author of all things—was pleased to work in that instance. On the hypothesis here brought forward, the *acarus Crossii* was a type of being ordained from the beginning, and destined to be realized under certain physical conditions. When a human hand brought these conditions into the proper arrangement, it did an act akin to hundreds of familiar ones which we execute every day, and which are followed by natural results; but it did nothing more. The production of the insect, if it did take place as assumed, was as clearly an act of the Almighty himself, as if he had fashioned it with hands. For the presumption that an act of aboriginal creation did take place, there is this to be said, that, in Mr. Weekes's experiment, every care that ingenuity could devise was taken to exclude the possibility of a development of the insects from ova. The wood of the frame was baked in a powerful heat; a bell-shaped glass covered the apparatus, and from this the atmosphere was excluded by the fumes constantly rising from the liquid, for the emission of which there was an aperture so arranged at the top of the glass, that only these fumes could pass. The water was distilled, and the substance of the silicate had been subjected to white heat. Thus every source of fallacy seemed to be shut up. In such circumstances, a candid mind, which sees nothing either impious or unphilosophical in the idea of a new creation, will be disposed to think that there is less difficulty in believing in such a creation having actually taken place, than in believing that, in two instances, separated in place and time, exactly the same insects should have chanced to arise from concealed ova.*

* See Appendix, H.

HYPOTHESIS OF THE DEVELOPMENT

OF THE

VEGETABLE AND ANIMAL KINGDOMS.

THE obvious gradation amongst the families of both the vegetable and animal kingdoms, from the simple lichen and animalcule respectively up to the highest order of dicotyledonous trees and mammalia, has already been intimated. Confining our attention, on this occasion, to the animal kingdom—it is to be observed that the gradation is much less simple and direct than is generally supposed. Even in its larger masses, it certainly does not proceed, at all parts of its course at least, upon one line; for the two sub-kingdoms of middle rank, mollusca and articulata, form unquestionably distinct approaches to the highest, the vertebrata. It even appears that there are intimations of more than two such great lines at various parts of the animal scale, as will be afterwards more particularly explained. Nevertheless, no doubt is entertained of a general advance of organization from the radiate into both the molluscous and articulate forms, and from the latter of these again (if not the former also) into the vertebrate; as also along the classes of (for example) the vertebrata, in this sequence—fishes, reptiles, birds, mammals.

While the external forms of all these various animals are so different, it is very remarkable that the whole are, after all, variations of a fundamental plan, which can be traced as a basis throughout the whole, the variations being merely modifications of that plan to suit the particular conditions in which each particular animal has been designed to live. Starting from the primeval germ, which, as we have seen, is the representative of a particular order of full-grown animals, we find all others to be merely advances from that type, with the extension of endowments and modification of forms which are required in each particular case; each form, also, retaining a strong affinity to that which precedes it, and tending to impress its own features on that which succeeds. This unity of structure, as it is called, becomes the more remarkable, when we observe that the organs, while preserving a resemblance, are often put to different uses. For example: the ribs become, in the serpent, organs of locomotion, and the snout is extended, in the elephant, into a prehensile instrument.

It is equally remarkable that identical purposes are served in different animals by organs essentially different. Thus, the mammalia breathe by lungs; the fishes, by gills. These are not modifications of one organ, but distinct organs. In mammifers, the gills exist and act at an early stage of the fœtal state, but afterwards go back and appear no more: while the lungs are developed. In fishes, again, the gills only are fully developed; while the lung structure either makes no advance at all, or only appears in the rudimentary form of an air-bladder. So, also, the baleen of the whale and the teeth of the land mammalia are different organs. The whale, in embryo, shows the rudiments of teeth; but these, not being wanted, are not developed, and the baleen is brought forward instead. The land animals, we may also be sure, have the rudiments of baleen in their organization. In many instances, a particular structure is found advanced to a certain point in a particular set of animals, (for instance, feet in the serpent tribe,) although it is not there required in any degree; but the peculiarity, being carried a little further forward, is perhaps useful in the next set of animals in the scale. In other instances, a portion of organization necessary in one sex is also presented in the other, where it is not necessary. For example, the mammæ of the human female, by whom these organs are obviously required, also exist in the male, who has no occasion for them. It might be supposed that in this case there was a regard to uniformity for mere appearance sake; but that no such principle is concerned, appears from a much more remarkable instance connected with the marsupial animals. The female of that tribe has a process of bone advancing from the pubes for the support of her pouch; and this also appears in the male marsupial, who has no pouch, and requires none. The rudimentary organs, as those not fully developed for use are called, appear most conspicuously in animals which form links between various families.

As formerly stated, the marsupials, standing at the bottom of the mammalia, show their affinity to the oviparous vertebrata, by the rudiments of two canals passing from near the anus to the external surfaces of the viscera, which are fully developed in fishes, being required by them for the respiration of aerated waters, but which are not needed by the atmosphere-breathing marsupials. We have also the peculiar form of the sternum and rib-bones of the lizards *represented* in the mammalia in certain white cartilaginous lines traceable among their abdominal muscles. The ostrich is an elevated form of the birds, approaching the mammalia, and in it we find the wings imperfectly or not at all developed, a diaphragm and urinary sac, (organs wanting in other birds,) and feathers resembling hair. Again, the ornithorhynchus is a mammal receding to near the grade of birds, and in it behold the bill and web-feet of the natatorial order of that class!

For further illustration, it is obvious that, various as may be the lengths of the upper part of the vertebral column in the mammalia, it always consists of the same parts. The giraffe has in its tall neck the same number of bones with the pig, which scarcely appears to have a neck at all.*

* D'Aubenton established the rule, that all the viviparous quadrupeds have seven vertebræ in the neck.

Man, again, has no tail; but the notion of a much-ridiculed philosopher of the last century is not altogether, as it happens, without foundation, for between the fifth and seventh week of the embryo a tail does exist, and in the mature subject the bones of this caudal appendage are found in a repressed state in the *os coccygis*. The limbs of all the vertebrate animals are, in like manner, on one plan, however various they may appear. In the hind-leg of a horse, for example, the angle called the hock is the same part which in us forms the heel; and the horse and other quadrupeds, with certain exceptions, walk, in reality, upon what answers to the toes of a human being. In this and many other quadrupeds the fore part of the extremities is shrunk up in a hoof, as the tail of the human being is shrunk up in the bony mass at the bottom of the back. The bat, on the other hand, has these parts largely developed. The membrane, commonly called its wing, is framed chiefly upon bones answering precisely to those of the human hand; its extinct analogue, the pterodactyle, had the same membrane extended upon the fore-finger only, which in that animal was prolonged to an extraordinary extent. In the paddles of the whale and other animals of its order, we see the same bones as in the more highly developed extremities of the land mammifers; and even the serpent tribes, which present no external appearance of such extremities, possess them in reality, but in an undeveloped or rudimental state.

The same law of development presides over the vegetable kingdom. Amongst phanerogamous plants, a certain number of organs are always present, either in a developed or rudimentary state; and those which are rudimentary can be developed by cultivation. The flowers which bear stamens on one stalk and pistils on another, can be caused to produce both, or to become perfect flowers, by having a sufficiency of nourishment supplied to them. So, also, where a special function is required for particular circumstances, nature provides for it, not by a new organ, but by a modification of a common one. Thus, for instance, some plants destined to live in arid situations, require to have a store of water which they may slowly absorb. The need is arranged for by a cup-like expansion round the stalk, in which water remains after a shower. Now the *pitcher*, as this is called, is not a new organ, but simply the metamorphosis of a leaf.

It thus appears, with regard to the constituent beings of large sections of the animal kingdom, that they are bound up in a fundamental unity, however various in degree of endowment and in the purposes which they serve in the world. They may be said to stand in a connexion analogous to that in which the planets are placed by the third law of Kepler. And the inference with regard to their origin is the same. Precisely as it is impossible to suppose a distinct exertion or fiat of Almighty Power for the formation of the earth, wrought up as it is in a complex dynamical connexion, first with Venus on the one hand and Mars on the other, and secondly with all the other members of the system, so is it impossible to conceive the same power using particular means for the production of a particular animal species, an individualized fraction, as it now appears, in a vast system which would not be complete without it, and into whose adjacent parts it melts by the finest shadings. Supposing, for a moment, that each species had been distinct in its origin, these shadings would have been unnecessary; and there would at least have been a strong probability against a unity of organization being assumed as part of the plan. In that case, abortive or rudimentary organs must have been considered as a kind of blemish—the thing of all others most irreconcileable with that idea of perfection which a general view of nature irresistibly attributes to its author. If, on the other hand, we admit that the animal kingdom took its rise in a general law, we see in the shadings and the organic unity something not only harmonious with, but essential to the system. Rudimentary organs, too, appear but as harmless peculiarities of development, and interesting evidences of the manner in which the Divine Author has been pleased to work.

We have yet to advert to the most interesting class of facts connected with our subject. First surmised by the illustrious Harvey, afterwards illustrated by Hunter in his wondrous collection at the Royal College of Surgeons, finally advanced to mature conclusions by Tiedemann, St. Hilaire, and Serres, embryotic development is now a science. Its primary positions are—1. that the embryos of all animals are not distinguishably different from each other; and, 2. that those of all animals pass through a series of phases of development, each of which is the type or analogue of the permanent configuration of tribes inferior to it in the scale. With regard to the latter proposition, it is to be remarked that, while it is generally true of the whole forms of animal being, it is more particularly true of departments of the organization, as the nutritive system, the vascular system, the nervous system, &c., each of which is destined for a peculiar degree of development in different groups of animals, according to their needs; and this, I may observe, is so far an explanation of the fact that a low class sometimes ascends in its highest forms to a point above the lowest forms of a class held on general grounds to be superior. Even in man there are some particulars of organization less developed than in certain animals which generally are far inferior. Speaking, however, roundly, it is undoubted, respecting nearly all animals, that they pass in embryo through phases resembling the general as well as the particular characters of those of lower grade. For example, the comatula, a free-swimming star-fish, is, at one stage of its early progress, a crinoid—that is, a star-fish fixed upon a stalk at the bottom of the sea. It advances from the form of one of the lower to that of one of the higher echinodermata. The animals of its first form were, as we have seen, among the most abundant in the earliest fossiliferous rocks: they began to decline in the New Red Sandstone era, and they were succeeded in the Oolitic age by animals *of the form of the mature comatula.* Thus, too, the insect, standing near the head of the articulated animals, is, in the larva state, an annelid or worm, the annelides being the lowest in the same class. Of the earth-worm, again, it has been observed that it passes through the forms of the polype, helianthois, and arenicola, before attaining its mature form. The higher crustacea, as the crab or lobster, at their escape from the ovum, resemble the perfect animal of the inferior order entomostraca, and pass through the forms of transition which characterize the intermediate tribes of crustacea. The salmon, a highly organized fish, exhibits, in its early stages, as has been remarked, the gelatinous dorsal cord, the heterocercal tail, and inferior position of the mouth, which mark the mature example of the cartilaginous fishes. The frog, again, for some time after its birth, is a fish with external gills, and other organs fitting it for an aquatic life, all of which are changed as it advances to maturity and becomes a land animal. The mammifer only passes through still more stages, according to its higher place in the scale. Nor is man himself exempt from this law. His first form is that which is permanent in the animalcule. His organization gradually passes through conditions generally resembling a fish, a reptile, a bird, and the lower mammalia, before it attains its specific maturity. At one of the last stages of his fœtal career, he exhibits an intermaxillary bone, which is characteristic of the perfect ape; this is suppressed, and he may then be said to take leave of the simial type, and become a true human creature. Even, as we shall find, the varieties of his race are represented in the progressive development of an individual of the highest, before we see the adult Caucasian, the highest point yet attained in the animal scale.

To come to particular points of the organization. The brain of man, which exceeds that of all other animals in complexity of organization and fulness of development, is, at one early period, only "a simple fold of nervous matter, with difficulty distinguishable into three parts, while a little tail-like prolongation towards the higher parts, and which had been the first to appear, is the only representation of a spinal marrow. Now, in this state it perfectly resembles the brain of an adult fish, thus assuming *in transitu* the form that in the fish is permanent. In a short time, however, the structure is become more complex, the parts more distinct, and the spinal marrow better marked; it is now the brain of a reptile. The change continues; by a singular motion, certain parts (*corpora quadrigemina*) which had hitherto appeared on the upper surface, now pass towards the lower; the former is their permanent situation in fishes and reptiles, the latter in birds and mammalia. This is another advance in the scale, but more remains yet to be done. The complication of the organ increases; cavities termed *ventricles* are formed, which do not exist in fishes, reptiles, or birds; curiously organized parts, such as the corpora striata, are added; it is now the brain of the mammalia. Its last and final change alone seems wanting, that which shall render it the brain of MAN."* And this change in time takes place.

So also with the heart. This organ, in the mammalia, consists of four cavities, but in the

* Lord's Popular Physiology.

reptiles of only three, and in fishes of two only, while in the articulated animals it is merely a prolonged tube. Now in the mammal fœtus, at a certain early stage, the organ has the form of a prolonged tube; and a human being may be said to have then the heart of an insect. Subsequently, it is shortened and widened, and becomes divided by a contraction into two parts, a ventricle and an auricle; it is now the heart of a fish. A subdivision of the auricle afterwards makes a triple-chambered form, as in the heart of the reptile tribes; lastly, the ventricle being also subdivided, it becomes a full mammal heart.

It is certainly very remarkable that, corresponding generally to these progressive forms in the development of individuals, has been the succession of animal forms in the course of time. Our earth bore crinoidea before it bore the higher echinodermata. It presented crustacea before it bore fishes, and when fishes came, the first forms were those cartilaginous types which correspond with the early fœtal condition of higher orders. Afterwards there were reptiles, then mammifers, and finally, as we know, came man. The tendency of all these illustrations is to make us look to *development* as the principle which has been immediately and mainly concerned in the peopling of this globe, a process extending over a vast space of time, but which is nevertheless connected in character with the briefer process by which an individual being is evoked from a simple germ. What mystery is there here—and how shall I proceed to enunciate the conception which I have ventured to form of what may prove to be its proper solution! It is an idea by no means calculated to impress by its greatness, or to puzzle by its profoundness. It is an idea more marked by simplicity than perhaps any other of those which have explained the great secrets of nature. But here, again, it may be said, lies one of its strongest claims to our faith.

My proposition is that the whole train of animated beings, from the simplest and oldest up to the highest and most recent, are the results, *first*, of an inherent impulse in the forms of life to advance, in definite times, through grades of organization terminating in the highest dicotyledons and mammalia; *second*, of external physical circumstances, operating reactively upon the central impulse to produce the requisite peculiarities of exterior organization,—the "adaptations" of the natural theologian. I contemplate these phenomena as ordained to take place in every situation, and at every time, where and when the requisite materials and conditions are presented —in other orbs as well as in this—in any geographical area of this globe which may at any time arise—observing only the variations due to difference of materials and of conditions. The nucleated vesicle is the fundamental form of all organization, the meeting-point between the inorganic and the organic—the end of the mineral and beginning of the vegetable and animal kingdoms, which thence start in different directions, but in a general parallelism and analogy. This nucleated vesicle is itself a type of mature and independent being, as well as the starting point of the fœtal progress of every higher individual in creation, both animal and vegetable. We have seen that it is a form of being which there is some reason to believe electric agency will produce—though not perhaps usher into full life—in albumen, one of those component materials of animal bodies, in whose combinations it is believed there is no chemical peculiarity forbidding their being any day realized in the laboratory. Remembering these things, we are drawn on to the supposition, that the first step in the creation of life, wherever it takes place, is *a chemico-electric operation, by which simple germinal vesicles are produced.* This is so much, but what are the next steps? I suggest, as an hypothesis countenanced by much that is ascertained, and likely to be further sanctioned by much that remains to be known, that the first step was *an advance under the impulse of the principle of development, from the simplest forms of being, to the next more complicated, and this through the medium of the ordinary process of generation.*

Unquestionably, what we ordinarily see of nature is calculated to impress a conviction that each species invariably produces its like. But I would here call attention to a remarkable illustration of natural law which has been brought forward by Mr. Babbage, in his *Ninth Bridgewater Treatise.* The reader is requested to suppose himself seated before the calculating machine, and observing it. It is moved by a weight, and there is a wheel which revolves through a small angle round its axis, at short intervals, presenting to his eye successively, a series of numbers engraved on its divided circumference.

Let the figures thus seen be the series, 1, 2, 3, 4, 5, &c., of natural numbers, each of which exceeds its immediate antecedent by unity.

"Now, reader," says Mr. Babbage, "let me ask you how long you will have counted before you are firmly convinced that the engine has been so adjusted, that it will continue, while its motion is maintained, to produce the same series of natural numbers? Some minds are so constituted, that after passing the first hundred terms, they will be satisfied that they are acquainted with the law. After seeing five hundred terms few will doubt, and after the fifty thousandth term the propensity to believe that the succeeding term will be fifty thousand and one, will be almost irresistible. That term *will* be fifty thousand and one; and the same regular succession will continue; the five millionth and the fifty millionth term will still appear in their expected order, and one unbroken chain of natural numbers will pass before your eyes, from *one* up to *one hundred million.*

"True to the vast induction which has been made, the next succeeding term will be one hundred million and one; but the next number presented by the rim of the wheel, instead of being one hundred million and two, is one hundred million *ten thousand* and two. The whole series from the commencement being thus,—

1
2
3
4
5
. . .
. . . .
. . . . .
. . . . . .
99,999,999
100,000,000

regularly as far as 100,000,001

100,010,002 the law changes.
100,030,003
100,060,004
100,100,005
100,150,006
100,210,007
100,280,008
... ...
... ...

"The law which seemed at first to govern this series failed at the hundred million and second term. This term is larger than we expected by

10,000. The next term is larger than was anticipated by 30,000, and the excess of each term above what we had expected forms the following table:—

10,000
30,000
60,000
100,000
150,000
. . . . . .
. . . . . .

being, in fact, the series of *triangular numbers*,* each multiplied by 10,000.

"If we now continue to observe the numbers presented by the wheel, we shall find, that for a hundred, or even for a thousand terms, they continue to follow the new law relating to the triangular numbers; but after watching them for

* The numbers 1, 3, 6, 10, 15, 21, 28, &c. are formed by adding the successive terms of the series of natural numbers thus:

$$1 = 1$$
$$1+2 = 3$$
$$1+2+3 = 6$$
$$1+2+3+4 = 10, \text{\&c.}$$

They are called triangular numbers, because a number of points corresponding to any term can always be placed in the form of a triangle; for instance:

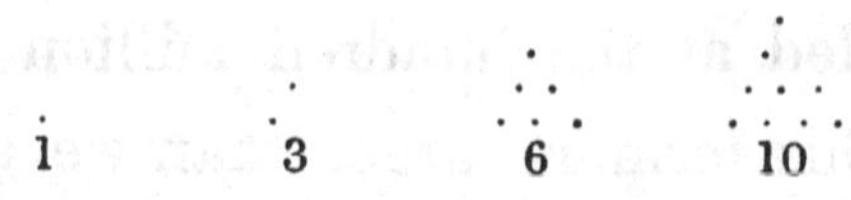

1 3 6 10

2761 terms, we find that this law fails in the case of the 2762nd term.

"If we continue to observe, we shall discover another law then coming into action, which also is dependent, but in a different manner, on triangular numbers. This will continue through about 1430 terms, when a new law is again introduced which extends over about 950 terms, and this, too, like all its predecessors, fails, and gives place to other laws, which appear at different intervals.

"Now it must be observed that *the law that each number presented by the engine is greater by unity than the preceding number*, which law the observer had deduced from an induction of a hundred million instances, *was not the true law that regulated its action*, and that the occurrence of the number 100,010,002 at the 100,000,002nd term was *as necessary a consequence of the original adjustment, and might have been as fully foreknown at the commencement, as was the regular succession of any one of the intermediate numbers to its immediate antecedent.* The same remark applies to the next apparent deviation from the new law, which was founded on an induction of 2761 terms, and also to the succeeding law, with this limitation only—that, whilst their consecutive introduction at various definite intervals, is a necessary consequence of the mechanical structure of the engine, our knowledge of analysis does not enable us to predict the periods themselves at which the more distant laws will be introduced."

It is not difficult to apply the philosophy of this passage to the question under consideration. It must be borne in mind that the gestation of a single organism is the work of but a few days, weeks, or months; but the gestation (so to speak) of a whole creation is a matter involving enormous spaces of time. Suppose that an ephemeron, hovering over a pool for its one April day of life, were capable of observing the fry of the frog in the water below. In its aged afternoon, having seen no change upon them for such a long time, it would be little qualified to conceive that the external branchiæ of these creatures were to decay, and be replaced by internal lungs, that feet were to be developed, the tail erased, and the animal then to become a denizen of the land. Precisely such may be our difficulty in conceiving that any of the species which people our earth is capable of advancing by generation to a higher type of being. Granting that, during the whole time which we call the historical era, the limits of species have been, to ordinary observation, and as far as the more conspicuous plants and animals are concerned, rigidly adhered to, we know the historical era to only a small portion of the entire age of our globe. We do not know what may have happened during the ages which preceded its commencement, as we do not know what may happen in ages yet in the distant future. All, therefore, that we can properly infer from the apparently invariable production of like by like is, that such is the ordinary procedure of nature in the time immediately passing before our eyes. Mr. Babbage's illustration shows how this ordinary procedure may be subordinate to a higher law which in proper season interrupts and changes it.

It has been seen that, in the reproduction of the higher animals, the new being passes through stages in which it is successively fish-like and reptile-like. But the resemblance is not to the adult fish or the adult reptile, but to the fish and reptile at a certain point in their fœtal progress; this holds true with regard to the vascular, nervous, and other systems alike. It seems as if gestation consisted of two distinct and independent stages—one devoted to the development of the new being through the conditions of the inferior types, or, rather, through the corresponding *first stages of their development;* another perfecting and bringing the new being to a healthy maturity, on the basis of the point of development reached. This may be illustrated by a simple diagram.* The fœtus of all the four classes may be supposed to advance in an identical condition to the point A. The fish there diverges and passes along a line apart, and peculiar to itself, to its mature state at F. The reptile, bird, and mammal, go on together to C, where the reptile diverges in like manner, and advances by itself to R. The bird diverges at D, and goes on to B. The mammal then goes forward in a straight line to the highest point of organization at M. This diagram shows only the main ramifications; but the reader must suppose minor ones, representing the subordinate differences of orders, families, genera, etc., if he wishes to extend his views to the whole varieties of being in the animal kingdom. Limiting ourselves at present to

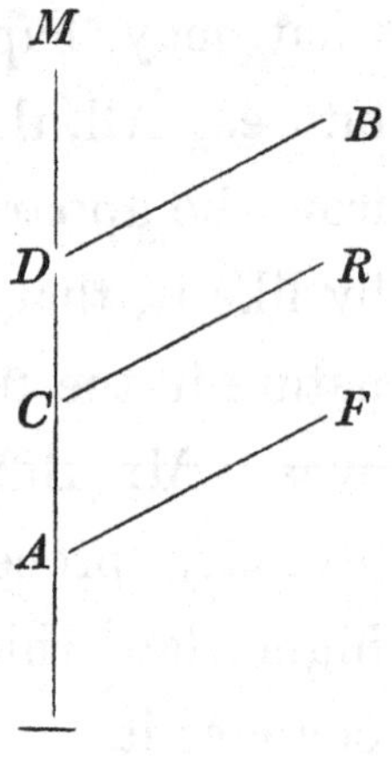

* Modified from one in Dr. Carpenter's Comparative Physiology.

the outline afforded by this diagram, it is apparent that the only thing required for an advance from one type to another in the generative process is that, for example, the fish embryo should not diverge at A, but go on to C before it diverges, in which case the progeny will be, not a fish, but a reptile. To protract the *straightforward part of the gestation over a small space* is all that is necessary.

To ascertain how this happened, is the great problem; but one not to be very readily solved, since our knowledge of nature is far from being clear on such subjects. The analogy, however, which we see between the embryotic progress of an individual of the highest class, and the progress of the animal kingdom from the protozoa of the early rocks up to the mammalia, strongly suggests that, just as a little vesicle, placed in the appropriate receptacle, advances in a few months to be a being of the highest existing organization,—urged on thereto by influences unseen, unknown, inconceivable, yet discernibly natural, and nothing more than natural,—so have the first and simplest of beings advanced upon their several lines of development through unguessed ages, till they at length became creatures of complicated and extensively-adapted structure,—this being similarly produced by an impulse, natural, though perhaps inscrutable, and equally liable to be affected by associated conditions, though applicable to a world and to centuries of centuries.

While we have here nothing but hypothesis to guide us, it is well to keep in view certain facts which seem to point strongly to the laws sought after. We still daily see organic development at work to certain effects only short of a transition from species to species. Sex is fully ascertained to be a matter of development. All beings are, at one stage of the embryotic progress, female; a certain number of them are afterwards *advanced* to be of the more powerful sex. From this it will be understood that no absolute distinction exists; all such are merely apparent. The ingenious Huber first made us aware of an instance, in a humble department of the animal world, of arrangements being made by the animals themselves for adjusting the law of development to the production of a particular sex. Amongst bees, as amongst several other insect tribes, there is in each community but one true female, the queen bee, the workers being false females or neuters; that is to say, sex is carried on in them to a point intermediate between the female and male, where it is attended by sterility. The preparatory states of the queen bee occupy sixteen days; those of the neuters, twenty; and those of males, twenty-four. Now it is a fact, settled by innumerable observations and experiments, that the bees can so modify a larva, which otherwise would result in a worker, that, when the perfect insect emerges from the pupa, it is found to be a queen or true female. For this purpose they enlarge its cell, make a pyramidal hollow to allow of its assuming a vertical instead of a horizontal position, keep it warmer than other larvæ are kept, and feed it with a peculiar kind of food. From these simple circumstances, leading to a *shortening* of the embryotic condition, results a creature different in form, and also in dispositions, from what would have otherwise been produced. Some of the organs possessed by the worker are here wanting. We have a creature " destined to enjoy love, to burn with jealousy and anger, to be incited to vengeance, and to pass her time without labour," instead of one " zealous for the good of the community, a defender of the public rights, enjoying an immunity from the stimulus of sexual appetite and the pains of parturition; laborious, industrious, patient, ingenious, skilful; incessantly engaged in the nurture of the young, in collecting honey and pollen, in elaborating wax, in constructing cells and the like!—paying the most respectful and assiduous attention to objects which, had its ovaries been developed, it would have hated and pursued with the most vindictive fury till it had destroyed them!"* All these changes may be produced by a mere modification of the embryotic progress, which it is within the power of the adult animals to effect. By the arrangements made and the food given, the embryo becomes sooner fit for being ushered forth in its imago or perfect state. Development may be said to be thus arrested at a particular stage—that early one at which the female sex is complete. In the other circumstances, it is allowed to go on four days longer, and a stage is then reached between the two sexes, which in this species is designed to be the perfect condition of a large portion of the community. Four days more make it a perfect male. It may be observed that there is, from the period of oviposition, a destined distinction between the sexes of the young bees. The queen lays the whole of the eggs which are designed to become workers, before she begins to lay those which be-

* Kirby and Spence.

come males. But the condition of her reproductive system evidently governs the matter of sex, for it is remarked that when her impregnation is delayed beyond the twenty-eighth day of her entire existence, she lays only eggs which become males.*

We have here, it will be admitted, a most remarkable illustration of the principle of development, although in an operation limited to the production of sex only. Let it not be said that the phenomena concerned in the generation of bees may be very different from those concerned in the reproduction of the higher animals. There is a unity throughout nature which makes the one case an instructive reflection of the other.

We shall now see an instance of development operating within the production of what approaches to the character of variety of species. It is fully established that a human family, tribe, or nation, is liable, in the course of generations, to be either

* M. Hampe has observed in the creeping willow (*salix repens*) that twigs above the water blossom as females, whilst those twigs which have been in the water, and subsequently blossomed when the water dried up, had only male blossoms. This seems a case analogous to that of the determination of sex by the bees, and may be held as an additional proof of the power of circumstances to affect development to very important results.

advanced from a mean form to a higher one, or degraded from a higher to a lower, by the influence of the physical conditions in which it lives. The coarse features, and other structural peculiarities of the negro race only continue while these people live amidst the circumstances usually associated with barbarism. In a more temperate clime, and higher social state, the face and figure become greatly refined. The few African nations which possess any civilization exhibit forms approaching the European; and when the same people in the United States of America have enjoyed a within-door life for several generations, they assimilate to the whites amongst whom they live. On the other hand, there are authentic instances of a people originally well-formed and good-looking, being brought, by imperfect diet and a variety of physical hardships, to a meaner form. It is remarkable that prominence of the jaws, a recession and diminution of the cranium, and an elongation and attenuation of the limbs, are peculiarities always produced by these miserable conditions, for they indicate an unequivocal retrogression towards the type of the lower animals. Thus we see nature alike willing to go back and to go forward. Both effects are simply the result of the operation of the law of development in the generative system.

Let us trace this law also in the production of certain classes of monstrosities. A human fœtus is often left with one of the most important parts of its frame imperfectly developed: the heart, for instance, goes no further than the three-chambered form, so that it is the heart of a reptile. There are even instances of this organ being left in the two-chambered or fish-form. Here we have apparently a realization of the converse of those conditions which carry on species to species, so far, at least, as one organ is concerned. Seeing a complete specific retrogression in one point, how easy it is to suppose a simply natural access of favourable conditions sufficient to reverse the phenomenon, and make a fish mother develop a reptile heart, or a reptile mother develop a mammal one. It is no great boldness to surmise that a super-adequacy in the measure of this under-adequacy (and the one thing seems as natural an occurrence as the other) would suffice in a natatorial bird to give its progeny the body of a rat, and produce the ornithorhynchus, or might give the progeny of an ornithorhynchus the mouth and feet of a true rodent, and thus complete at two stages the passage from the aves to the mammalia.

It may now be asked if scientific men are entitled to say that species is invariably persistent. They profess to have no instance on record of such a phenomenon as transmutation; but have they studied to observe, or treat fairly, facts of that character? The reader will judge upon this question, after a few particulars have been laid before him. The cowslip, primrose, oxlip, and polyanthus, have always been regarded by botanists as distinct species: therefore it might have been expected that each of these plants should be unalterable. It has lately, however, been found that the whole four can be changed into one another by culture. The artichoke of the garden and the cardoon (a kind of thistle) of the South American wild, are held as distinct species in all botanical works; yet it is found that the artichoke, in neglect, degenerates into the cardoon. The ranunculus aquatilis and the ranunculus hederaceus are, in like manner, set down as distinct species; but behold the secret of their difference! While the former plant remains in the water, its leaves are all finely cut and have their divisions hairy; but when the stems reach the surface, the leaves developed in the atmosphere are widened, rounded, and simply lobed. Should the seeds of this water plant fall upon a soil merely moist without being inundated, the result is the ranunculus hederaceus—the presumed distinct species—with short stalks, and none of the leaves divided into hairy cut work!* After such instances, it will not be surprising that the specific and even (so-called) generic differences among the cerealia are now discovered to be capable of reduction. It appears that, whenever oats sown at the usual time are kept cropped down during summer and autumn, and allowed to remain over the winter, a thin crop of rye is the harvest presented at the close of the ensuing summer.† Perhaps the greater number of what may be called the domesticated plants are unsuspected variations of others, which, growing wild, are recognised as different species. One noted instance of such transition has been detected within the last few years, in the common cabbage of the garden. This plant, with its stout fleshy stem and large succulent leaves gathered into a heart sometimes

* Lamarck, Philosophie Zoologique.

† Magazine of Nat. Hist., new series, i. 574. Gardener's Chronicle, Aug. and Sept. 1844.

reaching several feet in circumference, is now discovered to be merely an advance by means of external conditions from the wild kale of the sea-shore, which trails among the shingle with a tough slender stem and small glaucous leaf. After such an array of facts, can it reasonably be said that specific distinction is rigidly maintained in the current era?

The ready answer of opponents is, that in these cases, there has merely been a mistake in calling that species which was only variety. Species, say they, still stands as the appropriate term for the distinctive characters which never change in reproduction. This seems to be virtually a confession that the term is merely expressive of a hypothetical idea, that there is some point beyond which variation cannot be carried. And what is this but to present against numerous facts proving great variation—variation so great as to have long passed for specific—a limited experience showing other forms as usually persistent, while the same conditions are maintained? So far from there being anything settled on this point, there is not at the present day any characters generally agreed upon as determinative of species. It is admitted to be merely a term convenient in classification, but bearing no stamp from nature. Naturalists, in fact, know not the limits within which the variable properties of a race of animals is confined. They know not to what extent it will change under varied conditions. Looking to the case of a maritime plant transformed into a garden vegetable, or the ranunculus aquatilis into the ranunculus hederaceus, we may ask what more could be required, after dry land began to be formed, than that marine plants should extend their kind to the shore, in order to clothe the sun-lit earth with some of its earliest vegetable forms? Though we admitted that there are distinctions truly specific and not now, to ordinary observation, superable, we cannot say but that time may have a power even over these. Geology shows successions of forms, and grants enormous spaces of time within which we may believe them to have been changed from each other by the means which we see producing varieties. Brief spaces of time admittedly sufficing to produce these so-called varieties, is it unreasonable to suppose that large spaces of time would effect mutations somewhat more decided, but of the same character? And this more especially when we find that the era when these events took place was not a time of miracle in any other respect, but, on the contrary, one during which ordinary physical operations were going on exactly as they are doing at this day.

Seeing, then, that the alleged persistency of species is but a dubious objection, and finding, in the reproduction of one of the highest animals, the picture of a progress through the humbler forms,—this, moreover, being generally conformable to the actual history of Being presented by geology,—I do not see how, if natural causes are at all to be admitted in the case, we can stop short of a theory of Progressive Development, as the true explanation of the origin of organic nature. The idea which I form, and to which I trust to bring further support, is that the simplest and most primitive types, under a law to which that of like production is subordinate, gave birth to a type superior to it in compositeness of organization and endowment of faculties, that this again produced the next higher, and so on to the highest; the advance being in all cases small, but not of any determinate extent. There has been, in short, a universal gestation of nature, analogous to that of the individual being, and attended as little by circumstances of a startling or miraculous kind as the silent advance of an ordinary mother from one week to another of her pregnancy. We see but the chronicle of one or two great areas, within which the development has reached the highest forms. In some others, as Australia and the islands of the Pacific, development has not yet passed through the whole of its stages, because, owing to the comparatively late uprise of the land, the terrestrial portion of the development was there commenced more recently. It would commence and proceed in any new appropriate area, on this or any other sphere, exactly as it commenced upon our area in the time of the earliest fossiliferous rocks, whichever these are. Nay, it starts every hour with common infusions, and in similar humble theatres, and would there proceed through all the subsequent stages, granting suitable space and conditions. Thus simple—after ages of marvelling—is Organic Creation, and yet the whole phenomena are, in another point of view, wonders of the highest kind, being the undoubted results of ordinances arguing the highest attributes of foresight, skill, and goodness on the part of their Divine Author.*

Early in this century, M. Lamarck, one of the most distinguished of modern naturalists, suggested that the gradation of animals depended upon some general law which it was important

* See Appendix, I.

for us to discover. So far he was right; but the theory which he consequently formed with regard to the causes of the varieties of animated being was so far from being adequate to account for the facts, that it has had scarcely a single adherent. What M. Lamarck chiefly grounded upon was the well-known physiological fact, that use or exercise strengthens and enlarges an organ, while disuse equally atrophies it. He conceived that, an animal being brought into new circumstances, and called upon to accommodate itself to these, the exertions which it consequently made to that effect caused the rise of new parts: on the contrary, when new circumstances left certain existing parts unused, these parts gradually ceased to exist. Something analogous was, he thought, produced in vegetables, by changes in their nutrition, in their absorptions and transpirations, and in the quantity of caloric, light, air, and moisture which they received. This principle, with time, he deemed sufficient to have produced the advance from the monad to the mammal. His illustrations were chiefly of the following nature. The bird which is attracted to the water by the necessity of seeking there its food, wishes to move about on the surface of the flood, and for this purpose strikes out its toes. Through the consequent repeated separations of the toes, the skin uniting them at the roots is extended and at length becomes webbed. In like manner, the shore bird which has no desire to swim, but has to approach the water for food, is constantly subject to sink in the mud. The bird, disliking this, exerts all its efforts to lengthen its legs; the result is that, by continual habit for many generations, the legs of this order do at length become long and bare, as we see them. The error of the theory is in giving this adaptive principle too much to do. What undoubtedly is effectual in modifying the exterior peculiarities of animals was obviously insufficient to account for the great grades of organization. In the present day, we have superior light from geology and physiology, and hence arises my suggestion of a process analogous to ordinary gestation for advancing organic life through its grades in the course of a long but definite space of time, with only a recourse to external conditions as a means of producing the exterior characters. It must nevertheless be acknowledged that the germ of this natural view of the history of the world is presented in the work of Lamarck.

But the idea that any of the lower animals have been concerned in any way with the origin of man —is not this degrading? Degrading is a term expressive of a notion of the human mind, and the human mind is liable to prejudices which prevent its notions from being invariably correct. Were we acquainted for the first time with the circumstances attending the production of an individual of our race, we might equally think them degrading, and be eager to deny them, and exclude them from the admitted truths of nature. Knowing this fact familiarly and beyond contradiction, a healthy and natural mind finds no difficulty in regarding it complacently. So also, on becoming aware of the genetic history of our species, we might expect a rational and well-ordered mind to receive the idea with submission, as a view of the manner in which Divine Providence has been pleased in this instance to work. One source of the prejudice here to be contended with rests in our associations with the word ancestry. From seeing our immediate seniors possessed of venerable qualities, we naturally incline to venerate an ancestry; we presume its constituent elements to be something superior to ourselves. When called upon, therefore, to place any of the inferior orders of Being in this relation, a shock unavoidably follows. But here the error lies in transferring our idea of the qualities of a sire or grandsire to a collective ancestry. The elder people of the earth are in reality its children, and we are its true senate. The feeling due to early generations is the half-pitying benevolence which we daily bestow upon childhood. It follows that the still earlier generations antecedent to the perfection of the human type, ought to be regarded with an extension of this same feeling—the modification of it which humane natures daily exemplify in their treatment of the inferior animals. Our children, it may be said, are the representatives of the first simple and impulsive men of the earth: the lower animals represent the earlier pre-human stages of life. The right conception of the case, then, is that, in these stages, we are not to look for what is venerable, but, on the contrary, for what is crude and elementary. We are to expect but the *primitiæ* of man's masterful life—something not even ascending to the dignity of "the infant mewling in its nurse's arms." If thus prepared, we should experience no shock on hearing that the human form was preceded genealogically by others of humbler aspect,—no more than we are on learning that every individual amongst us passes through the characters of the invertebrate, fish, and reptile, before he is permitted to breathe the breath of life. A deep moral principle seems involved in the history of the origin of man. He is the undoubted chief of all creatures, and as such may well have a character and destiny in some respects peculiar and infinitely exalted above the rest; but it appears that his relation to them is, after all, one of kindred. Along with his authority over them, he bears from nature an obligation to abstain from wantonly injuring them, and as far as possible to cherish and protect them. Good men feel this duty, as if it were a command from a source above themselves. It seems to them that, if the helplessness of childhood calls for kind and gentle treatment, much more does the essentially weaker character of the dumb creature. And if the innocence of infancy is touching, still more so is the even more harmless character which (overlooking carnivorous instincts implanted for a wise purpose) attaches to the lower animals. It is common, under the influence of prejudice, to do gross injustice to the characters of these denizens of nature's common. We do not sufficiently reflect on their respectable qualities. Yet we must go to the dog for a type of the virtue of fidelity, and to the bee for that of industry. The parental affection of many animals is not below, if it is not considerably above, that of human mothers. Man nowhere exemplifies the virtue of patience in the practical perfection in which we see it in the horse and many other creatures which become the slaves of his convenience. Nowhere does he display that perfect moderation in wants. Alas for man's boasted superiority—in how many respects does it fail beside the unassuming merits of the mere commonalty of nature!

AFFINITIES AND GEOGRAPHICAL DISTRIBUTION OF ORGANISMS.

Plants and animals are classed into families, orders, classes, and sub-kingdoms, according as they agree in a smaller or greater number of characters, and take rank in compositeness or complexity of organization. The first difficulty of naturalists is to attain to a natural mode of classification, for there ever is a tendency to proceed upon certain formal, artificial ideas of our own, as presuming that the Divine Author of nature has acted upon some plan involving merely human ideas of regularity. Other difficulties arise when we inquire by what properties various organisms should be classed, many of these being identical or parallel in plants or animals, which in other respects are very different. It cannot be said that, in either kingdom, a satisfactorily natural system has been attained, nor, I venture to say, will any such system ever be formed, until there shall be a general consent to take a natural origin of organisms as partly our guide.

It is in the meantime seen with sufficient clearness, that the families of the vegetable and animal kingdoms can be arranged in grades from the highest forms in both instances to certain humble and simple types, amongst which it is hardly possible to discern any difference. In other words, both kingdoms start from certain humble and almost common forms, and advance to types of high organization in their respective departments. Dismissing, for convenience' sake, the vegetable kingdom, as in all essential respects a reflection of the other, and sure to fall into any natural scheme which may be approved of for its counterpart,—we see the animal kingdom dividing itself into four great sections, to which Cuvier has given the names, Radiata, Articulata, Mollusca, and Vertebrata; the first being the humblest, and the last the most exalted. Gradation has been observed in this kingdom from an early period; it is seen when we view comprehensive groups in succession, and also when we take these groups and analyse them singly; but it has always hitherto failed when an attempt was made to connect the whole into one line or simple scale. Some groups of animals—as, for example, the Arachnida (mites and spiders)—stand altogether apart, though presenting characters which make their sub-kingdom a matter of no dubiety. Many, so far from joining on to the bottom of the group which on general grounds stand next above them, rise into much higher forms than the lowest of those superior groups. Even the most vague assignment of a grade is often attended with the difficulty, that a group of animals may be exalted in one department of its organization, and comparatively humble in another. The consequence is, that naturalists have now in general abandoned that ideal chain of being which has so long been a favourite notion, although the affinities between special groups continue to be everywhere admitted. The truth is, that the gradations extend along several lines, or chains, each comprehending a succession of organisms of one comprehensively peculiar character, and each possessing ramifications, the extremities of which have, as a matter of course, no affinity towards any other group. When thus viewed, the animal kingdom becomes perfectly intelligible as a range of connected being, such as we should expect from a system having *genealogy* for its basis.

The lines here spoken of are usually seen taking their commencement in the obscure and dubious sub-kingdom which Cuvier calls RADIATA, but which in reality is not composed exclusively of animals of rayed form, being rather a general receptacle for all creatures of simple structure, and whose organization and habits are as yet little known. Passing rapidly through this division, most of the lines advance into one or other of the higher sub-kingdoms: for one exception, we must cite the *entozoa*, (internal parasites,) a line having a peculiar theatre of origin and residence, which proceeds, in the family of cœlelmintha, to the borders of the articulate region, but there apparently stops, from there being no scope for further advance. It were absurd for naturalists to try to connect such a group of animals with any other, living, for example, in the sea, because there is no possibility of a genetic connexion between the two. All the animals ranked under radiata agree, however, by reason of their humble character, in requiring a fluid medium of existence. The great majority of them are marine (sponges, corals, starfishes, infusoria, etc.), and the sea is hence to be considered as the grand matrix of life upon our planet.

Two grand lines—by which is meant lines which may be composed of minor lines of cognate character—are distinguishable in the invertebrate animals, corresponding, in one large part of their course, to the mollusca and articulata of Cuvier; the first being characterized by an unsymmetrical organization depending on the importance of the nutritive system, (whence called by Owen, *Heterogangliata;*) the second, by symmetricality of structure, depending on the predominance of the nervous system, (hence called by the same anatomist, *Homogangliata.*)

The starting-point of the MOLLUSCA is in the infusorial animalcules, the *Monads*, or *Vibriones*, the *Polygastria*, and the *Rotifera;* a succession of miscroscopic, but gradually ascending forms, which become developed in this order in a common infusion of vegetable or animal matter. We see this line, or some of its constituent minor lines, proceeding by a series of easily recognisable steps to the *Bryozoa*, which again verge upon the Ascidians in the molluscous order of the *Tunicata.* That such is a true genetic series, is indicated by close affinity of forms; but such is not the only method of investigating the pedigrees of animals. A second method is to observe the series of forms which the various animals go through in embryo, as these invariably represent the animals lower in organization, and which, according to our hypothesis, were, in most instances, the progenitors of those forming the subject of inquiry. Taking this latter plan, we feel that we are receiving additional light respecting the pedigree of the Mollusca in general, when we learn that, "at their coming forth from the egg, they swim about for some time in a condition which can scarcely be termed animal; for there is not even a mouth leading to an internal cavity, nor are there any other organs of locomotion than the cilia, the action of which is involuntary." This description clearly points to the infusorial form.

The three lowest classes of Mollusca are thus ranged according to grade of organization—Tunicata, Brachiopoda, Lamellibranchiata: the two latter are the bivalve shell-fish of popular observation, headless, and mostly sessile, or destined to spend their lives in fixed positions. The *Tunicata* are similar in all essential respects, except in being of humbler organization, and enclosed, not in shells, but in a cartilaginous or coriaceous integument; whence their name. It thus appears that the *Brachiopoda,* which are the predominant fossils of the Lower Silurian era, are the first animals we meet with in this line, having parts capable of commemorating their existence. It is also interesting to find that, while the Brachiopoda are generally inhabitants of deep seas, the *Lamellibranchiata,* among which are included the oyster, muscle, and other testacea, affect the beds of shallow seas, whence they spread in a variety of genera, towards shores, the mouths of rivers, and into fresh water. This is a gradation in habitat, which we shall find to be particularly worthy of observation.

The three highest molluscan classes, univalved, possessing heads, and with hardly an exception destined for independent locomotion, stand apart from the bivalve orders; generally superior in organization, as beseems their higher destiny, but not on that account to be held as an advanced form in the same genealogy. The lowest univalve class—called *Pteropoda,* from their mode of progression by a couple of wing-like membranes projecting from the neck—may be described as marine slugs, generally of small size, many of them naked, others protected by a very delicate shell, which swim through the ocean in vast multitudes; one species (clio) being in such abundance in the circumpolar ocean as to form the chief food of the whale. Professor Edward Forbes expresses his opinion that the larva of the pteropod will yet most likely be found to resemble an ascidian polype; inferring a very brief descent from the starting-point of life in its class.

The *Gasteropoda*—a class of many families and genera, including limpets, whelks, cowries, snails, etc.—have comparatively a high organization, the nervous system more concentrated, the nutritive more elaborate, but yet are of sluggish habits, usually moving by alternate contractions and expansions of a fleshy disk placed upon their stomachs; hence the name of the class. Many of the gasteropods are naked, others possessed of but slender protection; some are destructives, but the general character is harmless. A clear gradation of forms passes through some of the families, from the simple cone of the limpet to the spiral of the snail. The descent of the order appears to be from some families of the preceding; for "they all," says a minute observer of nature,* "commence life under the same form, both of

* Professor Edward Forbes, in Jameson's Journal, xxxvi 326.

shell and animal; namely, a very simple spiral, helicoid shell, and an animal furnished with two ciliated wings or lobes, by which it can swim freely through the fluid in which it is contained. At this stage of the animal's existence, it corresponds to the permanent state of a Pteropod."

In the univalve mollusks, as in the bivalves, it clearly appears that the humblest families are destined to a fixed place in the depths of the ocean.

As we advance through the higher groups, we find, in parallel steps with an improvement in the organs of animal life (for example, the splitting of the sexes into different individuals), an advance in the sphere of existence, to a life on the surface of the ocean, to fresh water, and even to dry land. The humble *Helicidæ* (snails), a family of the Gasteropoda, are the first animals which we encounter as adventuring upon the firm surface of the globe. And it is interesting to remark, in this progression, the requisite change in the mode of respiration—namely, from branchiæ, the apparatus necessary in aquatic life, to a vascular air-sac, the first form of lungs—the proper breathing organ of terrestrial animals.

In the peculiarly destructive *Cephalopoda*, we

recognise the highest organization of which the heterogangliate form appears capable: it includes the orthoceratites, ammonites, belemnites, etc. of the early rock systems, and the nautilus and cuttlefish of the present era. Their descent is probably from the carnivorous families of the pteropoda; for "the nucleus of their shells," says the naturalist last quoted, "is a spiral univalve, similar in form to the undeveloped shells above alluded to [those of the embryo gasteropods]; and it is yet to be seen whether all cephalopoda do not commence their existence under a spiral-shelled pteropodous form." It has also been remarked, that "the shells of two species [of pteropoda] afford indications of a transition towards the cephalopoda; one resembling in its straight conical form the belemnite and many other extinct genera of that class, and the other having a partially formed chamber at the lower closed extremity; and similar evidence is afforded by their internal structure."* This genealogy, if it shall be affirmed, will afford an important illustration of the geological history, because it will show that cephalopoda might be expected to make their appearance as early in the rock series as

* Carpenter's General Physiology.

any other mollusks possessing parts equally fitted to commemorate their existence. These animals are to be supposed as an ultimate form, reached, not through the medium of all the lower molluscan orders, but only of one, and that one possessing hard parts of such delicacy as not to have more than a slight chance of preservation. And this, it may be remarked, would be in harmony with what we know of the economy of nature with respect to the destructive animals. They seem to bear a relation to those upon which they are destined to prey, and to be a necessary accompaniment to them. Hence they would require to be upon a different genetic line—which actually appears, in every advance of the animal kingdom, to be the case—and developed contemporaneously with the weaker tribes, the fertility of which would otherwise produce complete anarchy. Granting, then, this pedigree for the cephalopoda, it would be no anomaly in our theory, although remains of inferior mollusks should never be found lower down in any part of the earth.

The cephalopods, though so highly organized in comparison with the gasteropods, do not advance, like these, to land forms, with apparatus for aerial respiration. They are, as a class, restricted to a pelagic life, admitting of occasional appearances on the surface of the ocean. Their respiratory system is accordingly branchiate, yet with marks of grade which are worthy of observation. It is, with physiologists, a law determining animal grade, that "increased number [of parts] irrespective of correlative structure, in an organ of the animal body, is ever a mark of its inferiority."* By this test, the nautilus, with its four branchiæ, sinks below the cuttle-fish with only two; and such is the basis of a division of the cephalopoda. In the whole of this order, however, there is a remarkable advance of the nervous system, though only to the effect of enabling the animal to supply itself with food by conquest over the inferior tribes. The nervous centres, which in lower mollusca were only protected by coverings which also served to cover the rest of the body, now become of sufficient importance to have a special protection, in the form of cartilaginous plates, which naturalists interpret as the rudiment of an internal skeleton. In this way, the cephalopoda approach the borders of the vertebrate sub-kingdom.

The second grand line involves the ARTICULATA

* Professor Owen, Lectures on the Invertebrated Animals.

of Cuvier, or HOMOGANGLIATA of Owen; animals "composed of a succession of rings formed by the skin or outward integument, which from its hardness constitutes a kind of external skeleton."* Such at least is the general description; for the Annelides are naked animals, and in the Cirrhipoda we see "the homogangliate organization marked by a tegumentary testaceous coat of mail, which they seem to have borrowed from the molluscous type." That the articulate form is a grade in structure, is very clearly proved by the approach made to it by the entozoa, a series belonging to the lowest sub-kingdom. We do not, however, at present see any zoophytic forms by which we can say for certain that the articulata have been preceded in the way of genealogy. Perhaps these previous stages have been much fewer than would be supposed by one, who merely considers that there is a great number of animal families inferior in organization to the articulata. The maxim, *Natura per saltum nihil agit*, must be applied with some care that we do not deem that a leap which perhaps is none.

The necessity of taking liberal views of the

* Rymer Jones's Animal Kingdom, p. 184.

procedure of nature in animal organization is impressed upon us by a character found in the very first order of the articulata to which our attention is called. That the *Annelides* (worms) are the humblest of the articulate animals there is now no doubt; yet, unlike their superiors, almost all of them have *red blood*, a feature of the highest sub-kingdom. Four leading forms in this class are described. Of the *Tubicolidæ*, or those inhabiting tubes, the serpula is an example. It forms for its habitation, usually upon some sea-immersed stone, an irregularly twisted calcareous tube, out of which it presents, floating in the water, a fan-like branchial apparatus of beautiful colours. The second order, *Suctoria*, is represented by the well-known leech; the third by the earth-worm; the fourth by the sea-mouse (aphrodita). In all of these groups, we see distinct advances in organization, and this is traceable in some in an interesting conformity with changes of scene and mode of life, from fixed situations to free movement in the sea, from thence to the shore, and thence again to the land. From the Nais, a simple marine worm which at the recess of tide burrows in the sand, there is a clear passage to the common earth-worm, which adopts a similar retreat on land, and comes to the surface when rain is falling. The fourth order, *Dorsibranchiata,* so called because of gill tufts ranged along the back, have an equally clear affinity, implying ancestral relationship, to certain land animals, which, however, are regarded as an independent class. The nereis, a well known dorsibranchiate, is an animal of great length, composed of a consecutive series of rings, each having a couple of processes at each side, which are used as oars for propelling the body through the water. One species is four feet long, and consists of several hundred segments. By conversion of the water-breathing apparatus into one fitted for aerial respiration, an increase of firmness and density to the external integument, and the development of a couple of limbs for each ring of the body, we see the nereis, as it were, transmuted into the *Myriapod.** Here, however, there may be more than one line of passage; for the two great families of the myriapods, the Julidæ and Scolopendridæ, are diverse in character, the former being vegetable feeders, the latter carnivorous, and, as has been remarked, it appears as a rule in the genetic system, that true carnivores

* See the presumed steps of conversion fully described in Rymer Jones's Animal Kingdom, p. 224.

are always apart. Confining our view to the Scolopendridæ, we see a remarkable continuity of character and habits transmitted to them from the presumed marine ancestor, (nereis,) allowing for the altered medium of existence. The scolopendra is an animal furnished with powerful destructive organs; and, living under stones and the bark of trees, and in fissures generally, it is his custom to wind insidiously along, and dart upon any little animal which comes in his way. Of the nereides, on the other hand, we are told that they "usually live in the excavations of littoral rocks, in the hollows of sponges, in the interstices of the radicles of thalassiophytes, under stones, and in general in all bodies which present fissures more or less profound . . . They all appear to feed upon animal substances. . . M. Bosc tells us they live upon polypi and small worms, on which they throw themselves, by darting the anterior part of their body, which they have first contracted."

The next articulate class demanding attention is the *Crustacea*, animals in which the annular sections are covered with a calcareous shell, and provided with jointed limbs, the respiratory apparatus being branchial; all are aquatic, except some of the higher genera, which occasionally adventure upon the land. They are in two great groups, entomostraca and malacostraca, the former being the simpler, and exclusively marine. Emmerich considers the Trilobites which figure so conspicuously in the early rocks, as between the two divisions, but most nearly allied to the first; whence it would appear that the crustacea which make so early an appearance in the rock series, are humble animals, only preceded in their own sub-kingdom by a group, which, from their slight forms, might be ill-adapted for preservation in strata exposed after deposition to a high temperature. The crustaceous form rises through various grades to the *decapoda* (ten-footed), of which the lobster and crab are examples; and some of these, as is well known, for the most part have a terrestrial existence. As elsewhere remarked, the young of the decapoda are of the entomostracous form, and thus denote a passage of the one from the other.

The next class in general rank is the *Insecta*, a wonderfully varied group, yet all agreeing in having thirteen segments and three pairs of legs; all, moreover, respiring by means of tracheæ or tubes permeating the body,—an arrangement having reference to their peculiar mode of locomotion, which, in the majority of species, is by flight through the air. The fact of the greater number of insect genera passing, in their larva state, through the annelidan or myriapodous form, points to these classes as their genetic origin; yet this is a point on which the benefit of further investigation is desirable. In the case of the *Arachnida* (mites and spiders), the highest articulate class, no humbler form is traceable in the embryo; it is therefore impossible to assign them any pedigree. Can it be possible that the arachnida, or these with the insecta, have sprung almost or wholly at once from inorganic elements under the proper electric influences? On this subject, we are quite unprepared to make any positive affirmation; but it certainly is remarkable that in no department of the animal kingdom, besides the infusoria and entozoa, have there been more frequent appearances of an aboriginal commencement of life than in the insecta. The acarus, moreover, so often produced, from certain solutions, where ova were rigidly excluded, is a lowly member of the arachnida.

In the highly varied sub-kingdom radiata, a strong distinction is usually drawn in favour of one great class, the *echinodermata*, or star-fishes,—animals, in general, highly organized, and enjoying free movement at the bottom of the sea; also remarkable for the chain of affinities which passes from the lowest to the highest families, and which clearly appears to be also a chain of genetic connexion. The class, as far as traceable backwards, starts with the encrinus, or stone lily, a group of animals now nearly extinct, but of which many varieties flourished in the early seas. The creature consisted of a stomach and mouth, surrounded by long tentacles or arms, placed upon the top of a stalk fixed to the sea-bottom, the whole being composed of numberless minute calcareous plates, connected by gelatinous substance. In more advanced forms of the same order, (as the comatula and the extinct marsupite,) the body and arms desert the stalk, and betake themselves to a free-swimming life; but, as has been elsewhere mentioned, the young comatula lives for a time as an encrinus; that is, upon a stalk. Seeing that the same animal, in an earlier embryotic stage, represents a polypidom, we conclude that in the polypiaria is the origin of the echinodermatous line: it is first the polypidom, then the encrinus, then the free-swimming comatula, or feather star, the last being one of the most graceful animals in existence. In the higher genera of the latter family, the tentacles are shortened and reduced in number. In the *Ophiuræ*, there are only five long and simple rays projecting from the central body. Afterwards, in the *Asteriadæ*, or true star-fishes, the central part dilates step by step, until it fills up the interstices between the rays, and the form becomes a pentagonal disk. From this there is a clear passage to the *Echinus* or sea-urchin, which is merely a spheroidal animal in a calcareous case, through which numberless spines or tentacles project, for locomotion and the collection of food. This form again becomes elongated into the cylindrical soft-bodied *Holothuria*, with a circle of tentacles at the oral extremity; thence the transition is easy to the genus *Fistularidæ*, animals externally worm-like, and possessing the rudiment of a heart, with red blood in the arteries, so that, in this last echinoderm, we may be said to have come nearly, if not fully abreast with, the annelides, and to be approximating to some of the humbler fishes. The reader cannot fail to have been struck by the great number of forms passed through in this line, in comparison with any other, before leaving the radiate sub-kingdom; but, in reality, the echinodermata, though of radiated form, are much superior to the rest of that division in their organization, which is, if not complicated in the usual sense of naturalists, full of extremely curious minute work. Their whole destiny seems to be of a high kind, for in the stone record their line of forms stands parallel with others in which the whole of the three lowest sub-kingdoms are passed through. Polypiarian animals and encrinites appear in the Silurians and many subsequent formations: at the commencement of the carbonigenous era, the latter are so abundant, that we walk over large tracts of country, where the rocks beneath our feet are almost wholly composed of their remains. The Asteriadæ appear in the upper Silurians, and are but faintly seen until the Lias, when they become conspicuous. In the Oolite, the Echinidæ make their appearance. These are the last which we could expect to be preserved in rocks, as the higher families possess no hard parts; otherwise, we might perhaps have seen the succession continued into the holothuriæ and fistularidæ.

We are now done with the affinities of the three first sub-kingdoms. The arrangement must be held as liable to correction; but there appears no room to doubt that it is near to the truth, as it proceeds everywhere upon affinities which are admitted by all naturalists. Assuming its general correctness as a view of the genealogies of the three first sub-kingdoms, we learn from it that the divisions of Cuvier's and other classifications are rather descriptions of grades in organization, than truly distinct groups of animals. The justest divisions would represent genetic lines passing vertically through the whole or a portion of the grade classification. It also appears that these grades are reached by steps of various length in the succession of forms, and in various spaces of time, plainly denoting that the advances have followed no uniform rule, but have been in some measure obedient to external and incidental conditions.

We are now to trace the passage of the invertebrate into the lowest of the vertebrate classes,—PISCES. And here geology supplies us with a fact of great importance in favour of the development theory. The alliance of the cephalopoda to the *Cyclostoma*, a family of the cartilaginous fishes, is stated on the high authority of Professor Owen. In that family, says the learned anatomist, "eight free filaments are extended forwards from the circumference of the funnel-shaped orifice of the mouth, representing the eight ordinary arms of the Cephalopoda Dibranchiata, but arrested in their development because of the preponderating size of the caudal extremity of the body, which now forms the sole locomotive organ." That is to say, the cephalopodous molluscan form, with eight arms for locomotion and the catching of prey, is here seen converted or metamorphosed, the eight arms being reduced almost to extinction, but replaced by a development of tail, so that the animal becomes a fish, though one of a low kind.* Now the cyclostoma are a portion of the order of fishes which made its appearance in the Upper Silurian and Devonian era, following immediately upon a time when the cephalopoda were the predominant carnivorous form of marine life. The lower animal, when at a point of great abundance —that is, we may presume, when the vital forces of the class were in great strength—is succeeded by one in which we see a conversion of its form. Could there well be more satisfactory evidence of that plan of development which is required to explain the otherwise necessarily natural origin of the organic kingdoms? It is important also to

* See Appendix, Note J.

remark the progress from entirely soft animals, to an order bearing cartilaginous plates to protect a rudimental brain; from these, again, to an order having a skull and vertebral column of cartilage; a series of advances entirely conformable to phenomena seen in individual development. Nor is it to be overlooked that the presumed progeny exhibit, in their voracious character, and the functions they serve in nature, a perfect family likeness to their ancestry. The cartilaginous fishes were the chief police for keeping down the redundant life of the Devonian and Carboniferous seas, as the cephalopoda had been during the Lower and partly also during the Upper Silurian eras.

The approach made by the annelides to some of the humbler forms of fish indicates another passage from the invertebrate into the vertebrate animals; and this passage may have taken place in the Upper Silurian or Devonian era, as annelides are ascertained to have then existed. Perhaps some of the less destructive of the early cartilagines—the Lepidoids were such an inoffensive family—have had such an origin.

I would now suggest, as an inquiry worthy of the attention of geologists, whether the echinodermal line has not given rise to the more recently developed fish families,—those which enter upon the field in the cretaceous era. If the fistularidæ make, as appears, so near an approximation to the lowest bearers of the vertebratal type, I do not see how any preconceived ideas regarding the order of subkingdoms to be passed through should stand in the way, especially after so many traces of similar irregularity. The geological history of the animals in question is favourable to the conjecture, for the echinoderms are amongst the most conspicuous and important forms antecedent to the chalk era. Looking, indeed, at the enormous abundance of crinoidea in the carboniferous rocks, one can hardly avoid the idea that this peculiar form was destined for some important ultimate history. It might be suggested that the orders by which the fish class is thus entered are those placed by Cuvier at the bottom of the osseous fishes, the *Lophobranchi* and *Plectognathi*, which indicate their nearness to the invertebrate type by many features attaching to some or all of them, as imperfection and slow hardening of the skeleton, deficiency of ribs and fins, low and embryotic forms of mouth, dentition, and gills; the Lophobranchi, moreover, hatching their young in a pouch below the tail, after the manner of a family of animals equally low in the mammalia.

In the present state of this inquiry, it is impossible to give an entire genealogical tree of Being. Much must remain obscure and unindicated. Even of what is set forth, some parts, as has been remarked, must be held liable to correction under better light. Enough, however, is done for the present object, if such fragments of the great composite chain be shown, as afford proof that there is such a thing in nature, and that the idea of genetic succession of advancing forms is in harmony with it. In the Fish, we have one of the obscurer portions of the animal kingdom. The classifications of Cuvier and Agassiz are neither of them admitted to be *natural;* it is therefore not to be expected of a mere general student, that he should be able to display the class in all its genetic relations, however confident he may be, from what he sees elsewhere, that such relations exist. We find, however, three advances made to its lower confines from the invertebrate—namely, by the cephalopodous mollusks, by the annulose animals (annelides), and by the echinodermata; and perhaps there are others. And we see advances made in its upper confines to the next higher class, the REPTILIA, which succeed it in the strata and chronology of the earth, as in organization.

This great class presents at least five leading divisions or orders—*Batrachia* (frogs, toads, salamanders); *Chelonia* (turtles, tortoises); *Lacertilia* (lizards); *Sauria* (crocodiles, gavials); and *Ophidia* (snakes). To the cold blood of the fish, they add a higher circulatory organization, as also lungs for aerial respiration: all of them are oviparous. The embryotic history of the first order furnishes an illustration of the transmutation theory which has never been sufficiently regarded. The young are at first true fishes; in the course of a few weeks, they are transformed before our eyes into terrestrial reptiles. Let any one, then, who is at a loss to conceive the advance of species from the bottom to the top of the animal kingdom, only observe this clear example of transition from one class to another, and he must at once be satisfied not only of the possibility of the general fact, but of the means by which it is accomplished. It is remarkable of some of the batrachia (proteus, menobranchus, etc.) that they retain the branchiæ of the fish all their lives, and never leave the water. The lower chelonia are, in like manner, fitted for and devoted to aquatic life. And the same statement may be made regarding the ophidia and sauria.

The geological history of the reptiles harmonizes so far with this account of their organization. Towards the close of the Devonian age, after fishes had existed for the space of an entire formation, there arises a family assuming a trace of reptilian character, in an inner row of crocodilian teeth. The Sauroid fishes, as they are called, increase and multiply, and, several ages thereafter,—in the Muschelkalk,—the Enaliosauria, or fish-crocodiles (ichthyosaur, etc.), are presented, in which the passage to the reptile is clear and distinct. But, before this event in the saurian line, a similar and more effectual transition had taken place in at least two other animal series, resulting in those specimens of the lacertilian order which are found in the Keuper, and those batrachians upon which Mr. Owen has conferred the name of Labyrinthidonts. In these instances, our records are meagre, and it is therefore not surprising that specimens uniting the fish with the reptile, as is done by the enaliosauria, are not as yet found. But still the general affinity to the fish character, as well as a certain degree of aquatic habit, is shown in the biconcave vertebræ of these early lizards and frogs.

The next class above the reptiles is that of AVES, or birds, in which warm blood makes its first appearance, and which are marked by various other traits of superiority, particularly in the nervous system, though an oviparous mode of reproduction is still maintained. The birds differ from other vertebrate classes in a remarkable unity of structure; the variations in orders and families are of a far less essential kind, expressing indeed little more than external adaptations to the sphere and habits of life. It is by these adaptations that the birds have been classed; the principal groups being—*Natatores* or swimming birds (penguins, divers, auks, gulls, ducks, geese); *Grallatores* or wading birds (flamingoes, cranes, plovers); *Rasores*, or scraping birds (pheasants, domestic poultry, grouse); *Insessores* or perching birds; *Raptores*, or birds of prey. The unity in the organization of birds is a fact which we are called upon to keep strongly in view in seeking for the elements of their genetic history.

Lamarck pointed out that the three first orders of birds are manifestly inferior to the rest, in as far as their young are independent of nursing care immediately on leaving the shell. Of these again, the Natatores are allowed to be, generally, the humblest. This leads us to seek a starting-point for the bird class in the aquatic families. Amongst these we find the divers and penguins to be marked by peculiarly inferior characters, a very slight development of the anterior, and an unusually posterior position of the hinder extremities, the former being covered with feathers partaking of the nature of scales. Then the animal moves one or two hundred yards at a time under water, using its anterior extremities as oars or paddles. It never uses these as wings, and it hardly possesses the power of walking. The approach made, on the other hand, to these birds by the turtle, is admitted by all naturalists. Its mode of progression through the water is precisely that of the diver. There is even a species which has obtained a partially ornithic name—the hawk's bill turtle. A question arises—are the Chelonia the exclusive origin of the birds? To this idea the unity of organization in *Aves* lends some countenanee; but, on the other hand, the crocodile is allowed to make an approach to the birds, and the Rhynchosaur, a lacertian of the

New Red Sandstone, shows in its mouth and feet an advance to the characters of the same class. However this may be, it is at least important that the Chelonia give families of various habits answering to the leading distinctions in the class of birds. Many are herbivorous and gentle in character; others are as fierce and carnivorous as any bird of prey in existence.

The Natatorial Birds, which chiefly haunt the open sea, and only come ashore for incubation, may be said to be transformed into the Grallatorial, and thus fitted to seek their food in the shallow waters of shores and in marshy inland situations, by merely being raised upon longer legs and having an addition to the length of their bills, the feet at the same time losing much or all of the webbed form. From this last form, again, the Gallinaceous or Rasorial is another change, shortening and strengthening the limbs, neck, and bill, and adapting the animal to a vigorous walking life in dry inland regions and woods. Such is precisely the line of advancing habits which we have seen in other instances, and the affinities of the orders are perfectly in conformity with it. Limiting our view, in the meantime, to the special families from which the rasorial birds have been descended—we start with one line in the Divers, large aquatic birds haunting the northern seas; from it we advance to the Ducks of marine habits having the hind toe webbed, thence to Geese and Swans (as well as river ducks, with the hind toe free), whence again we pass to the Cranes (Gruidæ), true waders, but in whose bulky body, arching tail, and feet, there is a clear approach to the pheasants and fowls. So from the Grebes, a family associated with the divers, there is a passage by the Bustards (Otidæ) to the other leading rasorial form, the Tetraonidæ or Grouse. All of these families are distinguished by their comparatively vegetable diet and innocent character. All are birds of the northern hemisphere, so that there is no geographical objection to the supposed lines of genealogy. The specialties of a similar genesis of birds for the southern hemisphere may be left for future consideration.

We have to turn back again to the natatores for an origin of the purely carnivorous birds. Swimmers of that character are found in such species as the albatross and petrel, the former of which is one of the largest and most formidable animals of the whole class. From these we pass by such grallatorial forms as the secretary, to the falcons, eagles, vultures, and owls, the true raptores of naturalists, some of which, as the osprey, still retain an affection for the marine habitat. Here the grallatorial forms, it must be admitted, are much more scanty than in the other lines; yet, if we consider that the aquatic destructives are safe by virtue of their marine situation, and the others by the wild and inaccessible grounds which they haunt, we may see how rapacious waders, more exposed to man's enmity, might have long ago become for the most part extinct.

The chief order remaining for consideration—insessores—consists of birds of very various character. One great tribe—conirostres—is generally gentle and granivorous; another—dentirostres—is almost as remarkable for its predatory character as the falcons. In this order, we find the bulk greatly diminished, a peculiarity almost always seen in those families which are at the upper or most inland end of the various genetic lines. It therefore appears that the various insessorial tribes are merely advanced forms of those above described: the rasores pass by such families as the columbidæ (pigeons) to the *conirostres* (finches, starlings, crows, etc.); the raptorial birds are reduced into the *dentirostres*, so called from the toothed bill in which we see but a modification of that of the eagle, as in these birds we see only a limitation of destructive power in the ratio of abridged strength.

The remaining bird families are so few, that they must fall into the same historic plan as the rest. Only one calls for special attention—the *Cursores* or runners, including the ostrich and emeu. They are of great bulk and stature, approaching in some points of structure, as the stomach and feathers, to the mammalia. We must assign them an origin in the swimmers and waders, though the special ancestral families may not be pointed to. A grallatorial family advancing, not into grounds fit for the rasorial existence, but into vast sandy plains, such as are coursed by the ostrich, might, by the local influences, be deflected and advanced into this peculiar and somewhat anomalous character.

It is now important to remark that the bird families here, in consideration of organization and affinity, placed at the bottom of their class, are those which make their appearance earliest in the stone record. The Connecticut footmarks are of grallatorial, cursorial, and rasorial birds. The bird found by Dr. Mantell in the Wealden was grallatorial, supposed to belong to the family Ardeidæ (herons). It is equally remarkable, that there is no raptorial bird till we come to the Eocene Tertiary, and that only dubious traces of other orders appear before then, a fact harmonizing with the hypothesis of their later development.

Finally, we have to inquire into the connexions of the mammalia with these lower classes. The first glimpse of this grand type in the history of the globe is presented by the *Cetiosaur*, a huge reptile of the oolite, nearly allied to the marine Sauria, but exhibiting, in the form of the larger vertebræ, a clear affinity to the cetaceous character. Here, it appears, we have aquatic mammalia or whales, starting from the Enaliosauria. In the *Dinosauria,* huge land crocodiles of the oolite and Wealden, there is an equally clear approach, in certain bones, to the structure of the pachyderm mammalia. There is, however, an obscurity over the exact lines of connexion between the lower classes of vertebrata and the mammalia, in consequence apparently of the long blank in land zoology which is represented by the cretaceous formation. We see with tolerable certainty, that, as the fish connect with reptiles and these with birds, so do the reptiles and birds connect with mammalia,—an order which geology also approves;—thus the general fact of the continued development of animal life is placed beyond doubt. We also can discern tracings towards some of the points of junction. But, as stated, the exact connexions are wanting.

The *Monotremes* and *Marsupials* are mammals of low grade, making only a small advance from the bird type. In the former group, we see the Natator converted, in the ornithorhynchus, into a semi-rodent, and, in the echidna, into a semi-insectivore or mole, the webbed feet and bill being still present, and the brain birdlike. In the marsupials, the brain and various other parts of organization show a decided affinity to the feathered tribes. These are facts which no naturalist pretends to doubt. Possibly, however, this is a peculiar stage of animality which has only been developed in a limited portion of the earth's surface, while no such half-way advance has been made elsewhere. Considering the monotremes and marsupials in this light, we need not be surprised that amongst them are found representatives of so many orders of placental mammalia, as the rodents, insectivora, etc. We have in these only semi-mammalian *apices* of so many branches of distinct stirpes or lines of the animal kingdom.

Of the true or placental Mammalia, the *Rodentia*, *Edentata*, and *Insectivora*, are admitted by naturalists to exhibit, in many points of structure, affinities to the birds.

In several of the other orders, we start with inferiorly organized marine forms, the approach to which appears to have been from reptiles, also aquatic.

The relation of the *Cetacea* to the land pachyderms is admitted by many naturalists. It is seen in the thick and naked skin, the gigantic body, massive bones, bulky head, and even the variable and irregular teeth. Probably, the extinct *Dinotherium* is one of the connecting forms. The advanced development is chiefly seen in the extremities, which, after all, are still imperfect, in as far as the clumped metatarsus is a mean form. Afterwards, we find this improved to the digitigrade structure in the Equidæ. Here also is seen a corresponding advance in habitat, from the sea to the river bank and the jungle, and thence to dry inland regions. In another section of the cetes, that represented by the walrus, we apparently have the origin of the *Ruminantia.* The *Seals* are clearly the progenitors of the *Ursine* and *Musteline Carnivora,* there being a distinct gradation of forms towards those families, corresponding with an advance from aquatic to terrestrial habits. From a phocal origin have also proceeded the *Felinæ* (lion, panther, etc.); but here the littoral or paludinate form, as in the case of the raptorial birds, is probably extinct, either from similar causes, or because the uprise in the era of the post-tertiaries had destroyed their proper habitat and means of existence. The upper extremities of all those lines present animals usually of reduced size and tending to domestication. The Canidæ are such a termination to the ursine line,* as the house-cat is to the feline. An uprise of the body of the animal from a lurching plantigrade walk is another accompaniment usually seen in this advance to inland genera.

* It is a generally received principle with regard to hybridians, that the parent animals should be in some degree akin. The fact of the bear and dog having proved fruitful therefore appears as an additional evidence in favour of the somewhat startling pedigree assigned to the latter animal in the text. The near resemblance of the dentition is, however, the best testimony for this affiliation.

The highest order of Mammalia—to which I would suggest the comprehensive title of *Cheirotheria*, (handed animals,) and which includes bats, lemurs, and monkeys, as well as our own race—alone remains to be noticed. Its origin appears most likely to be found in the manatean section of the cetacea,—those innocuous and sociable animals which are found in several isolated parts of the earth; the manatee in the sea and rivers of Brazil, and in the Senegal river in Africa, the dugong in the Indian seas, and certain other species in the seas north-east from Asia. The external resemblance of these animals to the human type is so great, that many naturalists believe it to have been the source of the many reports respecting mermaids. The female sitting up in the water, and holding her young one by the flipper to her pectoral mammæ, strikingly recals the human mother. But the moral character of the genus is even more remarkable. Unlike the ruminants, which see a companion slain with indifference, the manatees cling around a wounded or captured associate, bewailing his fate, and making common cause with him. The *Simiadæ* (monkeys) are to be regarded as more immediate predecessors of the human genus,—although it is not improbable that the particular family from which it sprung no longer exists. It is worthy, however, of remark, that the seas and rivers haunted by the greater number of the manatean families are in close neighbourhood to the special districts where the Simiadæ abound, in America, Africa, and Asia. It may here be remarked that of all the reptilian orders, the batrachian is that which has best pretensions to a place in the origin of the Cheirotheria. "It is singular," says Dr. Roget, "that the frog, though so low in the scale of vertebrated animals, should bear a striking resemblance to the human conformation in its organs of progressive motion." It is the only animal besides man with a calf to its leg. It evidently "is making," says Dr. Roget, "an approximation to the higher orders of mammalia." The frog, however, is but a humble offshoot of the main line terminating in the Cheirotheria. There is something more like a lineal predecessor of the order in the Labyrinthidont of Owen, that massive batrachian, which leaves its hand-like footsteps in the New Red Sandstone, and then is seen no more. Not for nothing is it that we start at the picture of that strange impression,—ghost of anticipated humanity,—for apparently it really is so. In these things the superficial thinker will only see matter of ridicule: the large-hearted and truly devout man, who puts nothing of nature away from him, will, on the contrary, discover in them interesting traces of the ways of God to man, and a deeper breathing of the lesson, that whatever lives is to him kindred.

Enough, I trust, has now been done to show that the animal kingdom (and by analogy, the vegetable also) is composed of series of forms, in which affinities are ascertained in so many places, that they may be assumed in all, and that these, usually taking their origin in the radiate sub-kingdom, afterwards pass through higher grades, but not in every case through all, until the highest is reached. It appears that the grand matrix of organic being is the sea, that what may be called trunk lines pass through this medium as high as the mammalian type, and that the terrestrial families may all be regarded as branches of these marine lines, though in some instances a passage from one class form to another has taken place on land. Two principles are thus seen at work in the production of the organic tenants of the earth —first, a gestative development pressing on through the grades of organization, and bringing out particular organs necessary for new fields of existence; and, secondly, a variative power resting in external conditions, and working to minor effects, though these may sometimes be hardly distinguishable from the other. Everywhere along the central scale of organization, the land has been, as it were, a temptation or provocation to new and superior forms adapted for inhabiting it. We might almost regard the progression as the result of an aspiration towards new and superior fields of existence, as from the deep sea to the shallow or river-embouchure, from the shore to the bank, from that again to the higher ground in the interior. He may not yet be held as a very fanciful naturalist who would regard the megatherium as eager to climb the tree which he could only shake, and thus producing a progeny fitted to do that which was the object of his wishes,—or the rock-nose whale, which loves to rest its head on rocks beside the beach, as wishful of that mode of life which was at length vouchsafed to a more highly developed descendant. Such too may be found to be the true principle of perfectibility in nature—a continual, though it may be an irregularly shown tendency to press on to better and better powers,—an indefinite improveableness, which may work, as in seconds, in the individual, or strike hours in the species.

When the naturalists of modern times began to inquire into the geographical distribution of plants and animals, they quickly found that the prevalent notion of their dispersion from one common centre was untenable. From facts observed by them, they have latterly concluded that, so far from this being the case, there are many provinces of the earth's surface occupied by plants and animals almost wholly peculiar, and which must accordingly have had a separate origin. Professor Henslow, of Cambridge, speaks of no fewer than forty-five such provinces for the vegetable kingdom alone.

A botanical or zoological province is generally isolated in some manner,—either as an island in the midst of a wide ocean—as, for example, St. Helena or the Isle de Bourbon—or as a portion of a continent separated from the rest either by a range of high mountains, or by the boundaries of a climate. It is also found that elevation of position comes to the same effect with regard to vegetation as advance in latitude; so that, as we ascend a lofty mountain in a tropical country, we gradually pass through zones exhibiting the plants of kinds appropriate to temperate and arctic regions. Even the neighbourhood of a salt marsh, however isolated, exhibits plants appropriate to such a soil.

Fewer distinct zoological regions are enumerated, but perhaps only in consequence of imperfect observation. Here, however, the evidences against communication of organisms from one region to another are even more decided. If, however, it were surmised that the organisms of isolated regions had been communicated from other countries, and merely been modified in their new abodes, the disproof of the conjecture would be more positive with regard to the zoology of the question than the botany. For, while it might appear possible that seeds had been floated even five hundred miles to a new soil like that of the Isle de Bourbon, how can we account, by such a supposition, for the existence there of bats, reptiles, and other animals, the progenitors of which could never have swam four hundred and fifty miles for the sake of a change of residence? This island, be it remarked, is of volcanic origin, and known to have become dry land at a comparatively recent period.

The two great continents of the earth are the first zoological divisions of its surface. The animals as well as plants of the old and new world are specifically different, with very few exceptions; that is, they are different in the degree which naturalists agree to consider as sufficient to establish distinct species. But even North and South America present different animals. We also find that the animals in the north and south of Asia are different, and that most of the African species are distinct from those of Asia.

The differences are in some instances so great as to be held by naturalists as generic. Beyond this point, however, there are parities or identities. We see in all these various regions feline animals, ruminants, pachyderms, rodents, etc. Thus, for the lion and tiger of Asia, we have a different lion and the panther in Africa, the jaguar in South America, and the puma ranging from Brazil to Canada. Instead of the elk of Northern Europe and the argali of Siberia, we have, in North America, the moose deer and mountain sheep. Asia and Africa have elephants, to which the extinct mastodon of Northern Europe and the extinct mammoth of North America are parallels; and it now appears that even the horse, of which there are several varieties in the old world, was abundant in the new, at a period long antecedent to the introduction of the present breed by the colonists. Australia has its emeu, Africa its ostrich, and America her rhea, all similar animals, though specifically different. We find simiæ planted in three great regions—Southern Asia, Western Africa, and equinoctial America, but all of different character, those of America being peculiarly distinct in their want of the opposable thumb and of callosities in the seat, as also in the use of the tail as a prehensile instrument. Australia has no mammalian animals of her own besides the marsupials, which are represented by a few species in America; but to the southern part of the latter continent are confined the whole family of the sloths. Africa, in like manner, has exclusive possession of the giraffe. To North America belongs a great number of genera of birds quite peculiar to it, and also a greater number and variety of the rodents than are found in other parts of the earth. Similar facts could be stated respecting other classes of animals; but I limit attention to the mammalia, as being the most restricted in number and the best known.

Some principles governing the parity and variation of the organisms spread over different regions have been observed. It is found, for instance, that there is more uniformity between two continents in one hemisphere, than between two portions of one continent extending into different hemispheres. North America is zoologically more allied to Northern Europe than it is to South America. An island, however far apart, is apt to show zoological features reflective of those of the nearest continent. Two countries, again, divided only by a narrow sea, have usually the same flora.

Some principle affecting the development of the higher animals can also be detected, in connexion with geological chronology. Startling as it may appear, we are now assured that the present great continent comprising Europe, Asia, and Africa, has been, with minor changes in the relative position of sea and land, one theatre of organic being since the commencement of the existence of land animals upon the surface of the earth; that is to say, there has been, on one part or another of this geographical area, an uninterrupted chain of living forms from an early period in the secondary formation. This is the zoological province whose history is presented by the geologists; it is the oldest we are acquainted with. There are, however, some isolated regions which are known with certainty to have been in a condition of dry land for a less space of time. Such are the volcanic islands, of which the Isle of Bourbon is an example. Such also are the Galapagos islands, placed in the Pacific, above five hundred miles from South America. Now it is remarkable in such regions to find the mammalia either wholly wanting or in very small numbers.

Australia itself—a fifth great section of the habitable globe—appears to be one of these regions of an incomplete zoology. It is well known to have no native mammalia besides that humble implacental kind which are nearly peculiar to it. Professor Owen remarks how the fishes of the oolitic era—acrodus, psammodus, etc.—with the contemporary mollusks (trigoniæ and terebratulæ), which served these fishes for food, are represented in the living cestraceon which swims the Australian seas, with exactly the same sea mollusks to yield them sustenance. "Araucariæ and cycadeous plants likewise," he says, "flourish on the Australian continent, where marsupial quadrupeds abound, and thus appear to complete a picture of an ancient condition of the earth's surface, which has been superseded in our hemisphere by other strata and a higher type of mammalian organization." *

Such being the facts of the case, we are to inquire whether they best agree with an hypothesis of an origin of organisms by special Divine exertion, or that of their origination in Divine power working in the manner of natural law; and also, if the latter supposition appear preferable, how far the facts agree with the plan of animated nature delineated in the preceding pages.

It is remarkable at the very first that there is any variety of species in different regions, more especially as the species of one region usually thrive when transplanted to another of generally similar character in point of soil and climate. Had organisms been produced by special attention—taking this according to any ideas we can form of it—we might rather have expected to see identical plants in similar countries. It will not avail here to attribute the variation to the cultivation of variety as a principle on the part of the Divine Disposer, for the differences evidently follow no such principle, being of various intensities in near and in remote situations. In this

* British Fossil Mammalia and Birds, p. 69.

consideration, there is a great obstacle to the reception of the special-exertion hypothesis. It seems much more likely that organisms took their rise in germs springing from inorganic elements; which germs being different in accordance with such slight local differences in the combinations of the elements as physical studies inform us of, and the external conditions attending their development being also locally different, the resulting vessels of life were various accordingly. Such variations of result are exactly of a piece with hundreds of other simply natural events—for example, the difference of animals born at one birth; and similar natural causes are therefore presumable for them.

The facts respecting the geographical distribution of organisms are in perfect harmony with the plan of their origin, which, from the geological history, the principles of organic development, and their external affinities, has here been sketched. That plan *necessitates* the facts of distribution, which the other hypothesis does not. First, a development of vegetable organisms, we shall say, taking place in the sea, it is exactly what we would expect that they should spread upon the neighbouring shores in every direction, and that we should thus, for example, have one flora surrounding the Mediterranean. So it is also likely that islands should botanically and zoologically partake of the character of the neighbouring continents. In regions, on the other hand, sufficiently distant to be involved in the influence of diverse foci of life, we are to expect differences proportioned to the difference of original elements and also of conditions attending the development of the various lines; there we may only expect to see such ultimate parities attained as those between the emeu of Australia and the rhea of America, or the jaguar and puma of the latter continent and the tiger of Asia. Here it is important to observe that the cetacea and the marine birds in the neighbourhood of the different continents, present less variation than do the land mammals and birds: they have advanced less way along the lines, and have been less exposed to the conditions productive of external variations. In the case of a well-defined zoological region, such as the northern parts of North America, we see the indigenous animals expressly confined to those families which our plan sets forth as springing from the marine tribes above adverted to. There is the polar bear, with his various progeny, the brown bear, black bear, the wolf, fox, and dog; these from a phocal ancestry. The sea-otter, sprung from an allied stock, gives birth to the few musteline animals which dwell in these dreary regions. Then we have herbivorous cetes, giving rise to the moose deer and musk ox, these again being the progenitors of the goat and sheep. And finally we have the unusually numerous rodents from the aquatic birds, which nowhere are seen in greater numbers than on the borders of the Arctic Ocean. Such, with the mole, is the whole show of mammalia in this province: it is, it will be observed, of a limited kind; but it is interesting to remark that it presents nearly all the animals of that class, which we have supposed from their affinities to be descended from the marine families of which there is such abundance upon the adjacent ocean. And, supposing this ocean to be the *berceau* of these land animals, we can easily see why the terrestrial mammalia of Northern Europe should be more akin than those of South America. The Northern Ocean, spreading in one character of climate along the confines of the two first regions, enables a set of maritime animals which may have come into existence in any part of it, to spread into the two continents alike—the same bear, nearly the same ruminants, and so forth; but, if the Southern Ocean have possessed, as is likely from its distance, a different development of animal life from the Northern, and be supposed as sending off terrestrial animals in like manner into South America, the interposition of several great zones of different climate stands forth as a sufficient reason why there should not have been the same communication of zoological forms in that case to the hyperborean seas, as there was from those laving North America to those which dash upon Scandinavia, Russia, and Siberia.

The hypothesis is equally applicable to the imperfect developments of life upon the more recently raised lands, such as the Galapagos islands and Australia. Development is a matter of time, and in the case of these regions, the full time has not yet elapsed. It is therefore exactly what we might expect, upon the natural hypothesis, that animal life should have yet in them hardly reached the mammalian stage, the point which was attained in our elder and greater province about the time of the oolite.* But no

* See this argument more fully elucidated in *Explanations, a Sequel to the Vestiges, &c.*

rational cause for this imperfect zoological show can be presented in consonance with the plan of special exertions. Its advocates can only refer to some vague assumption regarding the Divine will, to which it is treason against judgment to come, while a single surmise of natural procedure remains unexhausted.

EARLY HISTORY OF MANKIND.

THE human race is known to consist of numerous nations, displaying considerable differences of external form and colour, and speaking in general different languages. This has been the case since the commencement of written record. It is also ascertained that the external peculiarities of particular nations do not change rapidly. While a people remain upon one geographical area, and under the influence of one set of conditions, they always exhibit a tendency to persistency of type, insomuch that a subordinate admixture of various type is usually obliterated in a few generations. Numerous as the varieties are, they have all been found classifiable under five leading ones:—1. The Caucasian, or Indo-European, which extends from India into Europe and Northern Africa; 2. The

Mongolian, which occupies Northern and Eastern Asia; 3. The Malayan, which extends from the Ultra-Gangetic Peninsula into the numerous islands of the South Seas and Pacific; 4. The Negro, chiefly confined to Africa; 5. The aboriginal American. Each of these is distinguished by certain general features of so marked a kind, as to suggest to many inquirers, that they have had distinct or independent origins. Of these peculiarities, colour is the most conspicuous: the Caucasians are generally white, the Mongolians yellow, the Negroes black, and the Americans red. The opposition of two of these in particular, white and black, is so striking, that of them, at least, it seems almost necessary to suppose separate origins. Of late years, however, the whole of this question has been subjected to a rigorous investigation by a British philosopher, who has been remarkably successful in adducing evidence that the human race might have had one origin, for anything that can be inferred from external peculiarities.

It appears from this inquiry,* that colour and other physiological characters are of a more superficial and accidental nature than was at one time

* See Dr. Prichard's Researches into the Physical History of Man.

supposed. One fact is at the very first extremely startling, that there are nations, such as the inhabitants of Hindostan, apparently one in descent, which nevertheless contain groups of people of almost all shades of colour, and likewise discrepant in other of those important features on which much stress has been laid. Some other facts, which may be stated in brief terms, are scarcely less remarkable. In Africa, there are Negro nations, —that is, nations of intensely black complexion, as the Jolofs, Mandingoes, and Kafirs, whose features and limbs are as elegant as those of the best European nations. While we have no proof of Negro races becoming white in the course of generations, the converse may be held as established, for there are Arab and Jewish families of ancient settlement in Northern Africa, who have become as black as the other inhabitants. There are also facts which seem to show the possibility of a natural transition by generation from the black to the white complexion, and from the white to the black. True whites (apart from Albinoes) are not unfrequently born among the Negroes, and the tendency to this singularity is transmitted in families. There is, at least, one authentic instance of a set of perfectly black children being born to an Arab couple, in whose ancestry no such blood had intermingled. This occurred in the valley of the Jordan, where it is remarkable that the Arab population in general have flatter features, darker skins, and coarser hair, than any other tribes of the same nation.*

The style of living is ascertained to have a powerful effect in modifying the human figure in the course of generations, and this even in its osseous structure. About two hundred years ago, a number of people were driven by a barbarous policy from the counties of Antrim and Down, in Ireland, towards the sea-coast, where they have ever since been settled, but in unusually miserable circumstances, even for Ireland; and the consequence is, that they exhibit peculiar features of the most repulsive kind, projecting jaws with large open mouths, depressed noses, high cheek bones, and bow legs, together with an extremely diminutive stature. These, with an abnormal slenderness of the limbs, are the outward marks of a low and barbarous condition all over the world; it is particularly seen in the Australian aborigines.

* Buckingham's Travels among the Arabs. This fact is the more valuable to the argument, as having been set down with no regard to any kind of hypothesis.

On the other hand, the beauty of the higher ranks in England is very remarkable, being, in the main, as clearly a result of good external conditions. "Coarse, unwholesome, and ill-prepared food," says Buffon, "makes the human race degenerate. All those people who live miserably are ugly and ill-made. Even in France, the country people are not so beautiful as those who live in towns; and I have often remarked that in those villages where the people are richer and better fed than in others, the men are likewise more handsome, and have better countenances." He might have added, that elegant and commodious dwellings, cleanly habits, comfortable clothing, and being exposed to the open air only as much as health requires, co-operate with food in increasing the elegance of a race of human beings.

Subject to these modifying agencies, and perhaps to some others of a less appreciable nature, connected with physical geography, there is, as has been said, a remarkable persistency in national features and forms, insomuch that a single individual thrown into a family different from himself is absorbed in it, and all trace of him lost after a few generations. Such permanency may, like that of species, be the rule, but the exceptive variations, which result from causes obvious or obscure, are also of a prominent character. They seem to tend most to occur among the humbler families of plants and animals, but also frequently take place in the very highest. A notable instance of variety-production in an animal family by no means low, is often referred to, as having taken place under the observation of persons still alive to attest it. On a New England farm there originated, in the latter part of the last century, a variety of sheep with unusually short legs, which was kept up by breeding, on account of the convenience in that country of having sheep which are unable to leap over low fences. The starting and maintaining a *breed* of cattle, that is, a variety marked by some desirable peculiarity, are familiar to a large class of persons. It appears only necessary, when a variety has been thus produced, that a union should take place between individuals similarly characterized, and that the conditions under which it has been produced should be persisted in, in order to establish it. Early in the last century, a man named Lambert was born in Suffolk, with semi-horny excrescences, of about half an inch long, thickly growing all over his body. The peculiarity was transmitted to his children, and was last heard of in a third generation. The peculiarity of six fingers on the hand and six toes on the feet, appears in like manner in families which have no record nor tradition of such a peculiarity having affected them at any former period, and it is then sometimes seen to descend through several generations. It was Mr. Lawrence's opinion, that a pair, in which both parties were so distinguished, might be the progenitors of a new variety of the race, who would be thus marked in all future time. We have but obscure notions of the laws which regulate this variability within specific limits; but we see them continually operating, and they are obviously favourable to the supposition that all the great families of men may have been of one stock.

The tendency of the modern study of the languages of nations is to the same point. The last fifty years have seen this study elevated to the character of a science, and the light which it throws upon the history of mankind is of a most remarkable nature.

Following a natural analogy, philologists have thrown the earth's languages into a kind of classification: a number bearing a considerable resemblance to each other, and in general geographically near, are styled a *group* or *sub-family;* several groups, again, are associated as a *family*, with regard to more general features of resemblance. Six families are spoken of.

The Indo-European family nearly coincides in geographical limits with those which have been assigned to that variety of mankind which generally shows a fair complexion, called the Caucasian variety. It may be said to commence in India, and thence to stretch through Persia into Europe, the whole of which it occupies, excepting Hungary, the Basque provinces of Spain, and Finland. Its sub-families are the Sanskrit, or ancient language of India, the Persian, the Slavonic, Celtic, Gothic, and Pelasgian. The Slavonic includes the modern languages of Russia and Poland. Under the Gothic, are (1) the Scandinavian tongues, the Norske, Swedish, and Danish; and (2) the Teutonic, to which belong the modern German, the Dutch, and our own Anglo-Saxon. I give the name of Pelasgian to the group scattered along the north shores of the Mediterranean, the Greek and Latin, including the modifications of the latter under the names of Italian, Spanish, &c. The Celtic was, from two to three thousand years ago, the speech of a considerable tribe dwelling in Western Europe; but these have since been driven before superior nations into a few corners, and are now only to be found in the highlands of Scotland, Ireland, Wales, Cornwall, and certain parts of France. The Gaelic of Scotland, Erse of Ireland, and the Welsh, are the only living branches of this sub-family of languages.

The resemblances amongst languages are of two kinds,—identity of words, and identity of grammatical forms; the latter being now generally considered as the most important towards the argument. When we inquire into the first kind of affinity among the languages of the Indo-European family, we are surprised at the great number of common terms which exist among them, and these referring to such primary ideas, as to leave no doubt of their having all been derived from a common source. Colonel Vans Kennedy presents nine hundred words common to the Sanskrit and other languages of the same family. In the Sanskrit and Persian, we find several which require no sort of translation to an English reader, as *pader*, *mader*, *sunu*, *dokhter*, *brader*, *mand*, *vidhava;* likewise *asthi*, a bone, (Greek, *osteon;*) *denta*, a tooth, (Latin, *dens*, *dentis;*) *eyeumen*, the eye; *brouwa*, the eye-brow, (German, *braue;*) *nasa*, the nose; *karu*, the hand, (Gr. *cheir*;) *genu*, the knee, (Lat. *genu;*) *ped*, the foot, (Lat. *pes*, *pedis;*) *hrti*, the heart; *jecur*, the liver, (Lat. *jecur;*) *stara*, a star; *gela*, cold, (Lat. *gelu*, ice;) *aghni*, fire, (Lat. *ignis;*) *dhara*, the earth, (Lat. *terra*, Gaelic, *tir;*) *arrivi*, a river: *nau*, a ship, (Gr. *naus*, Lat. *navis;*) *ghau*, a cow; *sarpam*, a serpent.

The inferences from these verbal coincidences were confirmed in a striking manner when Bopp and others investigated the grammatical structure of this family of languages. Dr. Wiseman pronounces that the great philologist just named, "by a minute and sagacious analysis of the Sanskrit verb, compared with the conjugational system of the other members of this family, left no doubt of their intimate and positive affinity." It was now discovered that the peculiar terminations or inflections by which persons are expressed throughout the verbs of nearly the whole of these languages, have their foundations in pronouns; the pronoun was simply placed at the end, and thus became an inflection. "By an analysis of the Sanskrit pronouns, the elements of those existing in all the other languages were cleared of their anomalies; the verb substantive, which in Latin is composed of fragments referable to two distinct roots, here found both existing in regular form; the Greek conjugations, with all their complicated machinery of middle voice, augments, and reduplications, were here found and illustrated in a variety of ways, which a few years ago would have appeared chimerical. Even our own language may sometimes receive light from the study of distant members of our family. Where, for instance, are we to seek for the root of our comparative *better?* Certainly not in its positive, good, nor in the Teutonic dialects in which the same anomaly exists. But in the Persian we have precisely the same comparative, *behter*, with exactly the same signification, regularly formed from its positive *beh*, good."*

The second great family of languages is the *Syro-Phœnician*, comprising the Hebrew, Syro-Chaldaic, Arabic, and Gheez or Abyssinian, being localized principally in the countries to the west

* Wiseman's Lectures on the Connexion between Science and Revealed Religion, i. 44. The Celtic has been established as a member or group of the Indo-European family, by the work of Dr. Prichard, *On the Eastern Origin of the Celtic Nations.* "First," says Dr. Wiseman, "he has examined the lexicon resemblances, and shown that the primary and most simple words are the same in both, as well as the numerals and elementary verbal roots. Then follows a minute analysis of the verb, directed to show its analogies with other languages, and they are such as manifest no casual coincidence, but an internal structure radically the same. The verb substantive, which is minutely analyzed, presents more striking analogies to the Persian verb than perhaps any other

and south of the Mediterranean. Beyond them, again, is the *African* family, which, as far as research has gone, seems to be in like manner marked by common features, both verbal and grammatical. The fourth is the *Polynesian* family, extending from Madagascar on the west, through the Indian Archipelago, besides taking in the Malayan dialect from the continent of India, and comprehending Australia and the islands of the western portion of the Pacific. This family, however, bears such an affinity to that next to be described, that Dr. Leyden and some others do not give it a distinct place as a family of languages.

The fifth family is the *Chinese*, embracing a large part of China, and most of the regions of Central and Northern Asia. The leading features of the Chinese language are, its consisting alto-

language of the family. But Celtic is not thus become a mere member of this confederacy, but has brought to it most important aid; for, from it alone can be satisfactorily explained some of the conjugational endings in the other languages. For instance, the third person plural of the Latin, Persian, Greek, and Sanscrit, ends in nt, nd, *ντι*, *ντο*, nti, or nt. Now, supposing, with most grammarians, that the inflections arose from the pronouns of the respective persons, it is only in Celtic that we find a pronoun that can explain this termination; for there, too, the same person ends in nt, and thus corresponds exactly, as do the others, with its pronoun, *hwynt*, or *ynt*.

gether of monosyllables, and being destitute of all grammatical forms, except certain arrangements and accentuations, which vary the sense of particular words. It is also deficient in some of the consonants most conspicuous in other languages, b, d, r, v, and z; so that this people can scarcely pronounce our speech in such a way as to be intelligible: for example, the word Christus they call *Kuliss-ut-oo-suh.* The Chinese, strange to say, though they early attained to a remarkable degree of civilization, and have preceded the Europeans in many of the most important inventions, have a language which resembles that of children, or deaf and dumb people. The sentence of short, simple, unconnected words, in which an infant amongst us attempts to express some of its wants and its ideas —the equally broken and difficult terms which the deaf and dumb express by signs, as the following passage of the Lord's Prayer:—"Our Father, heaven in, wish your name respect, wish your soul's kingdom providence arrive, wish your will do heaven earth equality," &c.—these are like the discourse of the refined people of the so-called Celestial Empire. An attempt was made by the Abbé Sicard to teach the deaf and dumb grammatical signs; but they persisted in restricting themselves to the simple signs of ideas, leaving the structure undetermined by any but the natural order of connexion. Such is exactly the condition of the Chinese language.

Crossing the Pacific, we come to the last great family in the languages of the aboriginal Americans, which have all of them features in common, proving them to constitute a group by themselves, without any regard to the very different degrees of civilization which these nations had attained at the time of the discovery. The common resemblance is in the grammatical structure as well as in words, and the grammatical structure of this family is of a very peculiar and complicated kind. The general character in this respect has caused the term Polysynthetic to be applied to the American languages. A long many-syllabled word is used by the rude Algonquins and Delawares to express a whole sentence: for example, a woman of the latter nation, playing with a little dog or cat, would perhaps be heard saying, "*kuligatschis*," meaning, "give me your pretty little paw;" the word, on examination, is found to be made up in this manner: *k*, the second personal pronoun; *uli*, part of the word wulet, pretty; *gat*, part of the word wichgat, signifying a leg or paw; *schis*, conveying the idea of littleness. In the same tongue, a youth is called *pilape*, a word compounded from the first part of pilsit, innocent, and the latter part of lenape, a man. Thus, it will be observed, a number of parts of words are taken and thrown together, by a process which has been happily termed *agglutination*, so as to form one word, conveying a complicated idea. There is also an elaborate system of inflection; in nouns, for instance, there is one kind of inflection to express the presence or absence of vitality, and another to express number. The genius of the language has been described as accumulative; it "tends rather to add syllables or letters, making farther distinctions in objects already before the mind, than to introduce new words."* Yet it has also been shown very distinctly, that these languages are based in words of one syllable, like those of the Chinese and Polynesian families; all the primary ideas are thus expressed: the elaborate system of inflection and agglutination is shown to be simply a further development of the language-forming principle, as it may be called—or the Chinese system may be described as an arrestment of this principle at a particular early point. It

* Schoolcraft.

has been fully shown, that between the structure of the American and other families, sufficient affinities exist to make a common origin or early connexion extremely likely. The verbal affinities are also very considerable. Humboldt says, "In eighty-three American languages examined by Messrs. Barton and Vater, one hundred and seventy words have been found, the roots of which appear to be the same; and it is easy to perceive that this analogy is not accidental, since it does not rest merely upon imitative harmony, or on that conformity of organs which produces almost a perfect identity in the first sounds articulated by children. Of these one hundred and seventy words which have this connexion, three-fifths resemble the Manchou, the Tongouse, the Mongol, and the Samoyed; and two-fifths, the Celtic and Tchoud, the Biscayan, the Coptic, and Congo languages. These words have been found by comparing the whole of the American languages with the whole of those of the Old World; for hitherto we are acquainted with no American idiom which seems to have an exclusive correspondence with any of the Asiatic, African, or European tongues.* Humboldt and others con-

* Views of the Cordilleras.

sidered these words as brought into America by recent immigrants; an idea resting on no proof, and which is much discountenanced by the common words being chiefly those which represent primary ideas; besides, we now know, what was not formerly perceived or admitted, that there are great affinities of structure also. I may here refer to a curious mathematical calculation by Dr. Thomas Young, to the effect, that if three words coincide in two different languages, it is ten to one they must be derived in both cases from some parent language, or introduced in some other manner. "Six words would give more," he says, "than seventeen hundred to one, and eight near 100,000, so that in these cases the evidence would be little short of absolute certainty." He instances the following words to show a connexion between the ancient Egyptian and the Biscayan:—

	BISCAYAN.	EGYPTIAN.
New	Beria	Beri.
A dog	Ora	Whor.
Little	Gutchi	Kudchi.
Bread	Ognia	Oik.
A wolf	Otgsa	Ounsh.
Seven	Shashpi	Shashf.

Now, as there are, according to Humboldt, one

hundred and seventy words common to the languages of the new and old continents, and many of these are expressive of the most primitive ideas, there is, by Dr. Young's calculation, overpowering proof of the original connexion of the American and other human families.

It seems to me, after a full consideration of this kind of evidence, in connexion with the development theory, that there is no reason to regard more than two local origins for the human race as *necessary*; namely, one for the Asiatic, American, and European varieties, and another for the African. The former seems to be connected with the great development of the quadrumana in southern Asia; the latter, with that of western Africa.

What is known of the migrations of the first group of races, and also their traditions, point to southern Asia as the scene of their origin. The lines of these migrations all converge, and are concentrated about the region of Hindostan. The language, religion, modes of reckoning time, and some other peculiar ideas of the Americans, are now believed to refer their origin to North-Eastern Asia. Trace them further back in the same direction, and we come to the north of India. The history of the Celts and Teutones represents them as coming from the east, the one after the other, successive waves of a tide of population flowing towards the north-west of Europe: this line being also traced back, rests finally at the same place. So does the line of Iranian population, which has peopled the east and south shores of the Mediterranean, Syria, Arabia, and Egypt. The Malay variety, again, rests its limit in one direction on the borders of India. Standing on that point, it is easy to see how this great section of the human family, originating there, might spread out in different directions, passing into varieties of aspect and of language as they spread, the Malay variety proceeding towards the Oceanic region, the Mongolians to the east and north, and sending off the red men as a sub-variety, the European population going off to the north-westward, and the Syrian, Arabian, and Egyptian, towards the countries which they are known to have so long occupied. The Negro alone is here unaccounted for; and that race is the one most likely to have had an independent origin, seeing that it is a type so peculiar in an inveterate black colour, and so humble in development. The traditions of the first section exhibit an agreement with this view of its origin. There is one among the Hindoos which places the cradle of the human family in Thibet; another makes Ceylon the residence of the first man.

It has of late years been a favourite notion with several writers, that the human race was at first in a highly civilized state, and that barbarism was a second condition. The principal argument for it is, that we see many examples of nations falling away from civilization into barbarism, while, in some regions of the earth, the history of which we do not clearly know, there are remains of works of art far superior to any which the present unenlightened inhabitants could have produced. It is to be readily admitted that such decadences are common: but do they necessarily prove that there has been anything like a regular and constant decline into the present state, from a state more generally refined? May not these be only instances of local failures and suppressions of the principle of civilization, where it had begun to take root amongst a people generally barbarous? This, at least, were as legitimate an inference from the facts which are known. But it is also alleged that we know of no such thing as civilization being ever self-originated. It is always seen to be imparted from one people to another. Hence, of course, we must infer that civilization at the first could only have been of supernatural origin. This argument appears to be founded on false premises, for civilization does sometimes rise in a manner clearly independent amongst a horde of people generally barbarous. A striking instance is described in the laborious work of Mr. Catlin on the North-American tribes. Far placed among those which inhabit the vast region of the north-west, and quite beyond the reach of any influence from the whites, he found a small tribe living in a fortified village, where they cultivated the arts of manufacture, realized comforts and luxuries, and had attained to a remarkable refinement of manners, insomuch as to be generally called "the polite and friendly Mandans." They were also more than usually elegant in their persons, and of every variety of complexion between that of their compatriots and a pure white. Up to the time of Mr. Catlin's visit, these people had been able to defend themselves and their possessions against the roving bands which surrounded them on all sides; but, soon after, they were attacked by small-pox, which cut them all off except a small party, whom their enemies rushed in upon and destroyed to a man. What is this but a repetition on a small scale of phenomena with which ancient history familiarizes us—a nation rising in arts and elegances amidst barbarous neighbours, but at length overpowered by the rude majority, leaving only a Tadmor or a Luxor as a monument of itself to beautify the waste? What can we suppose the nation which built Palenque and Copan to have been but only a kind of Mandan tribe, which chanced to have made its way further along the path of civilization and the arts, before the barbarians broke in upon it? The flame essayed to rise in many parts of the earth; but there were strong agencies working against it, and down it accordingly went, times without number; yet there was always a vitality in it, nevertheless, and a tendency to progress, and at length it seems to have attained a strength against which the powers of barbarism can never more prevail. The state of our knowledge of uncivilized nations is very apt to make us fall into error on this subject. They are generally supposed to be all at one point in barbarism, which is far from being the case, for in the midst of every great region of uncivilized men, such as North America, there are nations partially refined. The Jolofs, Mandingoes, and Kafirs, are African examples, where a natural and independent origin for the improvement which exists is as unavoidably to be presumed as in the case of the Mandans.

The most conclusive argument against the original civilization of mankind is to be found in the fact that we do not now see civilization existing anywhere except in certain conditions altogether different from any we can suppose to have existed at the commencement of our race. To have civilization, it is necessary that a people should be numerous and closely placed; that they should be fixed in their habitations, and safe from violent external and internal disturbance; that a considerable number of them should be exempt from the necessity of drudging for immediate subsistence. Feeling themselves at ease about the first necessities of their nature, including self-preservation, and daily subjected to that intellectual excitement which society produces, men begin to manifest what is called civilization; but never in rude and shelterless circumstances, or when widely scattered. Even civilized men, when transferred to a wide wilderness, where each has to work hard and isolatedly for the first requisites of life, soon show a retrogression to barbarism: witness the plains of Australia, as well as the backwoods of Canada and the prairies of Texas. Fixity of residence and thickening of population are perhaps the prime requisites for civilization, and hence it will be found that all civilizations as yet known have taken place in regions physically limited. That of Egypt arose in a narrow valley hemmed in by deserts on both sides. That of Greece took its rise in a small peninsula bounded on the only land side by mountains. Etruria and Rome were naturally limited regions. Civilizations have taken place at both the eastern and western extremities of the elder continent—China and Japan, on the one hand; Germany, Holland, Britain, France, on the other—while the great unmarked tract between contains nations decidedly less advanced. Why is this, but because the sea in both cases has imposed limits to further migration, and caused the population to settle and condense? —the conditions most necessary for social improvement.* Even the simple case of the Mandans affords an illustration of this principle, for Mr. Catlin expressly, though without the least regard

* The problem of Chinese civilization, such as it is—so puzzling when we consider that they are only, as will be presently seen, the child race of mankind—is solved when we look to geographical position producing fixity of residence and density of population.

to theory, attributes their improvement to the fact of their being a small tribe, obliged, by fear of their more numerous enemies, to *settle in a permanent village,* so fortified as to ensure their preservation. "By this means," says he, "they have advanced further in the arts of manufacture, and have supplied their lodges more abundantly with the comforts and even luxuries of life than any Indian nation I know of. The consequence of this," he adds, "is, that the tribe have taken many steps ahead of other tribes *in manners and refinements.*" These conditions can only be regarded as natural laws affecting civilization, and it might not be difficult, taking them into account, to predict of any newly settled country its social destiny. An island like Van Dieman's land might fairly be expected to go on more rapidly to good manners and sound institutions than a wide region like Australia. The United States might be expected to make no great way in civilization till they be fully peopled to the Pacific; and it might not be unreasonable to expect that, when that event has occurred, the greatest civilizations of that vast territory will be found in the peninsula of California and the narrow stripe of country beyond the Rocky Mountains. This, however, is a digression. To return: it is also necessary for a civilization that at least a portion of the community should be placed above mean and engrossing toils. Man's mind is subdued, like the dyer's hand, to that it works in. In rude and difficult circumstances, we unavoidably become rude, because then only the inferior and harsher faculties of our nature are called into exercise. When, on the contrary, there is leisure and abundance, the self-seeking and self-preserving instincts are allowed to rest, the gentler and more generous sentiments are evoked, and man becomes that courteous and chivalric being which he is found to be amongst the upper classes of almost all civilized countries. These, then, may be said to be the chief natural laws concerned in the moral phenomenon of civilization. If I am right in so considering them, it will of course be readily admitted that the earliest families of the human race, although they might be simple and innocent, could not have been in anything like a civilized state, seeing that the conditions necessary for that state could not have then existed. Let us only for a moment consider some of the things requisite for their being civilized,—namely, a set of elegant homes ready furnished for their reception, fields, ready cultivated to yield them food without labour, stores of luxurious appliances of all kinds, a complete social enginery for the securing of life and property,—and we shall turn from the whole conceit as one worthy only of the philosophers of Utopia.

Yet, as has been remarked, the earliest families might be simple and innocent, while at the same time unskilled and ignorant, and obliged to live merely upon such substances as they could readily procure. The traditions of all nations refer to such a state as that in which mankind were at first: perhaps it is not so much a tradition as an idea which the human mind naturally inclines to form respecting the fathers of the race; but nothing that we see of mankind absolutely forbids our entertaining this idea, while there are some considerations rather favourable to it. A few families, in a state of nature, living near each other, in a country supplying the means of livelihood abundantly, are generally simple and innocent; their instinctive and perceptive faculties are also apt to be very active, although the higher intellect may be dormant. If we therefore presume India to have been the cradle of our race, they might at first exemplify a kind of golden age;

but it could not be of long continuance The very first movements from the primal seat would be attended with deterioration, nor could there be any tendency to true civilization till groups had settled and thickened in particular seats physically limited.

The causes of the various external peculiarities of mankind now require some attention. Why, it is asked, are the Africans black, and generally marked by ungainly forms; why the flat features of the Chinese, and the comparatively well-formed figures of the Caucasians? Why the Mongolians generally yellow, the Americans red, and the Caucasians white? These questions were complete puzzles to all early writers; but physiology has lately thrown a great light upon them. It is now shown that the brain, after completing the series of animal transformations, passes through the characters in which it appears in the Negro, Malay, American, and Mongolian nations, and finally becomes Caucasian. The face partakes of these alterations. "One of the earliest points in which ossification commences is the lower jaw. This bone is consequently sooner completed than the other bones of the head, and acquires a predominance, which, as is well known, it never loses in the Negro. During the soft pliant state of the bones of the skull, the oblong form which they naturally assume, approaches nearly the permanent shape of the Americans. At birth, the flattened face, and broad smooth forehead of the infant, the position of the eyes rather towards the side of the head, and the widened space between, represent the Mongolian form; while it is only as the child advances to maturity, that the oval face, the arched forehead, and the marked features of the true Caucasian, become perfectly developed."* *The leading characters, in short, of the various races of mankind, are simply representations of particular stages in the development of the highest or Caucasian type.* The Negro exhibits permanently the imperfect brain, projecting lower jaw, and slender bent limbs, of a Caucasian child, some considerable time before the period of its birth. The aboriginal American represents the same child nearer birth. The Mongolian is an arrested infant newly born. And so forth. All this is as respects form;† but whence colour? This might be supposed to

* Lord's Popular Physiology, explaining observations by M. Serres.

† Conformably to this view, the beard, that peculiar attribute of maturity, is scanty in the Mongolian, and scarcely exists in the Americans and Negroes.

have depended on climatal agencies only; but it has been shown by overpowering evidence to be independent of these. In further considering the matter, we are met by the very remarkable fact that colour is deepest in the least perfectly developed type, next in the Malay, next in the American, next in the Mongolian, the very order in which the degrees of development are ranged. *May not colour, then, depend upon development also?* We do not, indeed, see that a Caucasian fœtus at the stage which the African represents is anything like black; neither is a Caucasian child yellow, like the Mongolian. But the case of a Caucasian fœtus, or child, at any of its stages of development, is different from that of a being whose *mature form* only comes up to the same point. When a being is presented, who at full time has only attained a point of formation such as the Caucasian passed at a comparatively early stage of this embryotic history, there may be a character of skin liable to a certain tinting on being exposed. Development being arrested at so immature a stage in the case of the Negro, the skin may take on the colour as an unavoidable consequence of its imperfect organization. It is favourable to this view, that Negro infants are not deeply black at first, but only acquire the full colour after exposure for some time to the atmosphere; also that the parts of the body concealed by clothing are not generally of so deep a hue as the face and hands. The phenomenon, in short, appears identical in character with the photographic process; not a result of the action of heat, as has been so long blunderingly supposed, but of light! It takes its place under the infant science of actino-chemistry, to which, perhaps, many other remarkable phenomena connected with the natural history of our race will yet be referred. This view, seeming to account for all the varieties of mankind as only the result of so many gradations in the developing power of the human mothers, is favourable to the doctrine of one origin; but it cannot be considered as settling the question, seeing that separate developments may have attained various points in the scale of the human organization—as one of the pachydermatous lines has reached the full equine form in Asia, but only the comparatively humble quagga in Africa.

We have seen that the traces of a common origin in all languages afford a ground of presumption for the unity of at least the principal portion of the human race. They establish a still stronger probability that that portion of mankind had not yet begun to disperse before they were possessed of a means of communicating their ideas by conventional sounds — in short, speech. This is a gift so peculiar to man, and in itself so remarkable, that there is a great inclination to surmise a miraculous origin for it, although there is no proper ground, or even support, for such an idea in Scripture, while it is clearly opposed to everything else we know with regard to the providential arrangements for the creation of our race. Here, as in other cases, a little observation of nature might have saved much vain discussion. The real character of language itself has not been thoroughly understood. Language, in its most comprehensive sense, is the communication of ideas by whatever means. Ideas can be communicated by looks, gestures, and signs of various other kinds, as well as by speech. The inferior animals possess some of those means of communicating ideas, and they have likewise a silent and unobservable mode of their own, the nature of which is a complete mystery to us, though we are assured of its reality by its effects. Now, as the inferior animals were all in being before man, there was language upon earth long ere the history of our race commenced. The only additional fact in the history of language, which was produced by our creation, was the rise of a new mode of expression—namely, that by *sound-signs* produced by the vocal organs. In other words, speech was the only novelty in this respect attending the creation of the human race. No doubt it was an addition of great importance, for, in comparison with it, the other natural modes of communicating ideas are insignificant. Still, the main and fundamental phenomenon, language, as the communication of ideas, was no new gift of the Creator to man; and in speech itself, when we judge of it as a natural fact, we see only a result of some of those superior endowments of which so many others have fallen to our lot through the medium of a superior organization.

The first and most obvious natural endowment concerned in speech is that peculiar organization of the larynx, trachea, and mouth, which enables us to produce the various sounds required. Man started at first with this organization ready for use, a constitution of the atmosphere adapted for the sounds which that organization was calculated to produce, and, lastly, but not leastly, as will afterwards be more particularly shown, a mental power within, prompting to, and giving directions for, the expression of ideas. Such an arrangement of mutually adapted things was as likely to produce sounds as an Eolian harp placed in a draught is to produce tones. It was unavoidable that human beings so organized, and in such a relation to external nature, should utter sounds, and also come to attach to these conventional meanings, thus forming the elements of spoken language. The great difficulty which has been felt is to account for man going in this respect beyond the inferior animals. There could have been no such difficulty if speculators in this class of subjects had looked into physiology for an account of the superior vocal organization of man, and had they possessed a true science of mind to show man possessing a faculty for the expression of ideas which is only rudimental in the lower animals. Another difficulty has been in the consideration that, if men were at first utterly untutored and barbarous, they could scarcely be in a condition to form or employ language—an instrument which it requires the fullest powers of thought to analyze and speculate upon. But this difficulty also vanishes upon reflection—for, in the first place, we are not bound to suppose the fathers of our race as early attaining to great proficiency in language, and, in the second, language itself seems to be amongst the things least difficult to be acquired, if we can form any judgment from what we see in children, most of whom have, by three years of age, while their information and judgment are still as nothing, mastered and familiarized themselves with a quantity of words, infinitely exceeding in proportion what they acquire in the course of any subsequent similar portion of time.

Discussions as to which parts of speech were first formed, and the processes by which grammatical structure and inflections took their rise, appear in a great measure needless, after the matter has been placed in this light. The mental powers could readily connect particular arbitrary sounds with particular ideas, whether those ideas were nouns, verbs, or interjections. As the words of all languages can be traced back into roots which are monosyllables, we may presume these sounds to have all been monosyllabic accordingly. The clustering of two or more together to express a compound idea, and the formation of inflections by additional syllables expressive of pronouns and such prepositions as of, by, and to, are processes which would or might occur as matters of course, being simple results of a mental power called into action, and partly directed, by external necessities. This power, however, as we find it in very different degrees of endowment in individuals, so would it be in different degrees of endowment in nations, or branches of the human family. Hence we find the formation of words and the process of their composition and grammatical arrangement, in very different stages of development in different races. The Chinese have a language composed of a limited number of monosyllables, which they multiply in use by mere variations of accent, and which they have never yet attained the power of clustering or inflecting; the language of this immense nation—the third part of the human race—may be said to be in the condition of infancy. The aboriginal Americans, so inferior in civilization, have, on the other hand, a language of the most elaborately composite kind, perhaps even exceeding, in this respect, the languages of the most refined European nations. These are but a few out of many facts tending to show that language is in a great measure independent of civilization, as far as its advance and development are concerned. Do they not also help to prove that cultivated intellect is not necessary for the origination of language?

Facts daily presented to our observation afford equally simple reasons for the almost infinite diversification of language. It is invariably found that, wherever society is at once dense and refined, language tends to be uniform throughout the whole population, and to undergo few changes in the course of time. Wherever, on the contrary, we have a scattered and barbarous people, we have great diversities, and comparatively rapid alterations of language; insomuch that, while English, French, and German, are each spoken with little variation by many millions, there are islands in the Indian archipelago, probably not inhabited by one million, but in which there are hundreds of languages, as diverse as are English, French, and German. It is easy to see how this should be. There are peculiarities in the vocal organization of every person, tending to produce peculiarities of pronunciation: for example, it has been stated that each child in a family of six gave the monosyllable, fly, in a different manner, (eye, fy, ly, &c.) until, when the organs were more advanced, correct example induced the proper pronunciation of this and similar words. Such departures from orthoepy are only to be checked by the power of example; but this is a power not always present, or not always of sufficient strength. The self-devoted Robert Moffat, in his work on South Africa, states, without the least regard to hypothesis, that amongst the people of the towns of that great region, "the purity and harmony of language is kept up by their pitchos or public meetings, by their festivals and ceremonies, as well as by their songs and their constant intercourse. With the isolated villages of the desert, it is far otherwise. They have no such meetings; they are compelled to traverse the wilds, often to a great distance from their native village. On such occasions, fathers and mothers, and all who can bear a burden, often set out for weeks at a time, and leave their children to the care of two or three infirm old people. The infant progeny, some of whom are beginning to lisp, while others can just master a whole sentence, and those still further advanced, romping and playing together, the children of nature, through the live-long day, *become habituated to a language of their own.* The more voluble condescend to the less precocious, and thus, from this infant Babel, proceeds a dialect composed of a host of mongrel words and phrases, joined together without rule, and *in the course of a generation the entire character of the language is changed.*" * I have been told, that in like manner the children of the Manchester factory workers, left for a great part of the day, in large assemblages, under the care of perhaps a single elderly person, and spending the time in amusements, are found to make a great deal of new language. I have seen children in other circumstances amuse themselves by concocting and throwing into the family circulation entirely new words; and I believe I am running little risk of contradiction when I say that there is scarcely a family, even amongst the middle classes of this country, who have not some peculiarities of pronunciation and syntax, which have originated amongst themselves, it is hardly possible to say how. All these things being considered, it is easy to understand how mankind have come at length to possess between three and four thousand languages, all different at least as much as French, German, and English, though, as has been shown, resemblances suggesting a common origin are observable in most of them.

What has been said on the question whether mankind were originally barbarous or civilized,

* Missionary Scenes and Labours in South Africa.

will have prepared the reader for understanding how the arts and sciences, and the rudiments of civilization itself, took their rise amongst men. The only source of fallacious views on this subject is the so frequent observation of arts, sciences, and social modes, forms, and ideas, being not indigenous where we see them now flourishing, but known to have been derived elsewhere: thus Rome borrowed from Greece, Greece from Egypt, and Egypt itself, lost in the mists of historic antiquity, is now supposed to have obtained the light of knowledge from some still earlier scene of intellectual culture. This has caused to many a great difficulty in supposing a natural or spontaneous origin for civilization and the attendant arts. But, in the first place, several stages of derivation are no conclusive argument against there having been an originality at some earlier stage. In the second, such observers have not looked far enough, for, if they had, they might have seen various instances of civilizations which it is impossible, with any plausibility, to trace back to a common origin with others; such are those of China and America. They would also have seen civilization springing up, as it were, like oases amongst the arid plains of barbarism, as in the case of the Mandans. A still more attentive study of the subject would have shown, amongst living men, the very psychological procedure on which the origination of civilization and the arts and sciences depended.

These things, like language, are simply the effects of the spontaneous working of certain mental faculties, each in relation to the things of the external world on which it was intended by creative Providence to be exercised. The monkeys themselves, without instruction from any quarter, learn to use sticks in fighting, and some build houses—an act which cannot in their case be considered as one of instinct, but of intelligence. Such being the case, there is no necessary difficulty in supposing how man, with his superior mental organization, (a brain five times heavier,) was able, in his primitive state, without instruction, to turn many things in nature to his use, and commence, in short, the circle of the domestic arts. He appears, in the most unfavourable circumstances, to be able to provide himself with some sort of dwelling, to make weapons, and to practise some simple kind of cookery. But, granting, it will be said, that he can go thus far, how does he ever proceed further unprompted, seeing that many nations remain fixed for ever at this point, and seem unable to take one step in advance? It is perfectly true that there is such a fixation in many nations; but, on the other hand, all nations are not alike in mental organization, and another point has been established, that only when some favourable circumstances have settled a people in one place, do arts and social arrangements get leave to flourish. If we were to limit our view to humbly endowed nations, or the common class of minds in those called civilized, we should see absolutely no conceivable power for the origination of new ideas and devices. But let us look at the inventive class of minds which stand out amongst their fellows—the men, who, with little prompting or none, conceive new ideas in science, arts, morals—and we can be at no loss to understand how and whence have arisen the elements of that civilization which history traces from country to country throughout the course of centuries. See a Pascal, reproducing the Alexandrian's problems at fifteen; a Ferguson, making clocks from the suggestions of his own brain, while tending cattle on a Morayshire heath; a boy Lawrence, in an inn on the Bath road, producing, without a master, drawings which the educated could not but admire; or look at Solon and Confucius, devising sage laws, and breathing the accents of all but divine wisdom, for their barbarous fellow-countrymen, three thousand years ago—and the whole mystery is solved at once. Amongst the arrangements of Providence is one for the production of original, inventive, and aspiring minds, which, when circumstances are not decidedly unfavourable, strike out new ideas for the benefit of their fellow-creatures, or put upon them a lasting impress of their own superior sentiments. Nations, improved by these means, become in turn *foci* for the diffusion of light over the adjacent regions of barbarism—their very passions helping to this end, for nothing can be more clear, than that ambitious aggression has led to the civilization of many countries. Such is the process which seems to form the destined means for bringing mankind from the darkness of barbarism to the day of knowledge and mechanical and social improvement. Even the noble art of letters is but, as Dr. Adam Fergusson has remarked, "a natural produce of the human mind, which will rise spontaneously, wherever men are happily placed;" original alike amongst the ancient Egyptians and the dimly monumented Toltecans of Yucatan. "Banish," says Dr. Gall, "music, poetry, painting, sculpture, architecture, all the arts and sciences, and let your Homers, Raphaels, Michael Angelos, Glucks, and Canovas, be forgotten, yet let men of genius of every description spring up, and poetry, music, painting, architecture, sculpture, and all the arts and sciences, will again shine out in all their glory. Twice within the records of history has the human race traversed the great circle of its entire destiny, and twice has the rudeness of barbarism been followed by a higher degree of refinement. It is a great mistake to suppose one people to have proceeded from another on account of their conformity of manners, customs, and arts. The swallow of Paris builds its nest like the swallow of Vienna, but does it thence follow that the former sprung from the latter? With the same causes we have the same effects; with the same organization we have the manifestation of the same powers."

MENTAL CONSTITUTION OF ANIMALS.

No clear ideas have as yet been entertained by the generality of even educated men, with regard to the mental constitution of animals. The very nature of this constitution is not as yet generally known or held as ascertained. There is, indeed, a notion of old standing, that the mind is in some way connected with the brain; but the metaphysicians insist that it is, in reality, known only by its acts or effects, and they accordingly present the subject in a form which is unlike any other kind of science, for it does not so much as pretend to have a basis in nature. There is a general disinclination to regard mind in connexion with organization, from a fear that this must interfere with the cherished religious doctrine of the spirit of man, and lower him to the level of the brutes. A distinction is therefore drawn between our mental manifestations and those of the lower animals, the latter being comprehended under the term instinct, while ours are collectively described as mind, mind being again a received synonyme with soul, the immortal part of man. There is here a strange system of confusion and error, which it is most imprudent to regard as essential to religion, since candid investigations of nature tend to show its untenableness. There is, in reality, nothing to prevent our regarding man as being specially, in accordance with his position as the head or chief of all animals, endowed with an immortal spirit, at the same time that his ordinary mental manifestations are looked upon as simple phenomena resulting from organization, those of the lower animals being phenomena absolutely the same in character, though developed within narrower limits.*

* "Is not God the first cause of matter as well as of mind? Do not the first attributes of matter lie as inscrutable in the bosom of God—of its first author—as those of mind? Has not even matter confessedly received from God the power of experiencing, in consequence of impressions from the earlier modifications of matter, certain consciousnesses called sensations of the same? Is not, therefore, the wonder of matter also receiving the consciousnesses of other matter called ideas of the mind a

What has chiefly tended to take mind, in the eyes of learned and unlearned, out of the range of nature, is its apparently irregular and wayward character. How different the manifestations in different beings! how unstable in all!—at one time so calm, at another so wild and impulsive!

wonder more flowing out of and in analogy with all former wonders, than would be, on the contrary, the wonder of this faculty of the mind not flowing out of any faculties of matter? Is it not a wonder which, so far from destroying our hopes of immortality, can establish that doctrine on a train of inferences and inductions more firmly established and more connected with each other than the former belief can be, as soon as we have proved that matter is not perishable, but is only liable to successive combinations and decombinations?

"Can we look further back one way into the first origin of matter than we can look forward the other way into the last developments of mind? Can we say that God has not in matter itself laid the seeds of every faculty of mind, rather than that he has made the first principle of mind entirely distinct from that of matter? Cannot the first cause of all we see and know have *fraught matter itself, from its very beginning, with all the attributes necessary to develop into mind*, as well as he can have from the first made the attributes of mind wholly different from those of matter, only in order afterwards, by an imperceptible and incomprehensible link, to join the two together?

" * * [The decombination of the matter on which mind rests] is this a reason why mind must be annihilated? Is the temporary reverting of the mind, and of the sense out of which that mind develops, to their original component elements, a reason for thinking that they cannot again at another later period

It seemed impossible that anything so subtle and aberrant could be part of a system, the main features of which are regularity and precision. But the irregularity of mental phenomena is only in appearance. When we give up the individual, and take the mass, we find as much uniformity of result as in any other class of natural phenomena.

and in another higher globe, be again recombined, and with more splendour than before? * * The New Testament does not, after death here, promise us a soul hereafter unconnected with matter, and which has no connexion with our present mind—a soul independent of time and space. That is a fanciful idea, not founded on its expressions, when taken in their just and real meaning. On the contrary, it promises us a mind like the present, founded on time and space; since it is, like the present, to hold a certain situation in time, and a certain locality in space; but it promises a mind situated in portions of time and of space different from the present: a mind composed of elements of matter more extended, more perfect, and more glorious: a mind which, formed of materials supplied by different globes, is consequently able to see further into the past, and to think further into the future, than any mind here existing: a mind which, freed from the partial and uneven combination incidental to it on this globe, will be exempt from the changes for evil to which, on the present globe, mind as well as matter is liable, and will only thenceforth experience the changes for the better which matter, more justly poised, will alone continue to experience: a mind which, no longer fearing the death, the total decomposition, to which it is subject on this globe, will thenceforth continue last and immortal."—HOPE, *on the Origin and Prospects of Man*, 1831.

The irregularity is exactly of the same kind as that of the weather. No man can say what may be the weather of to-morrow; but the quantity of rain which falls in any particular place in any five years is precisely the same as the quantity which falls in any other five years at the same place. Thus, while it is absolutely impossible to predict of any one Frenchman that during next year he will commit a crime, it is quite certain that about one in every six hundred and fifty of the French people will do so, because in past years the proportion has generally been about that amount, the tendencies to crime in relation to the temptations being everywhere invariable over a sufficiently wide range of time. So also, the number of persons taken in charge by the police in London for being drunk and disorderly in the streets, is, week by week, a nearly uniform quantity, showing that the inclination to drink to excess is always in the mass about the same, regard being had to the existing temptations or stimulations to this vice. Even mistakes and oversights are of regular recurrence, for it is found in the post-offices of large cities, that the number of letters put in without addresses is year by year the same. Statistics has ascertained an equally distinct regularity in a wide range, with regard to many other things concerning the mind, and the doctrine founded upon it has lately produced a scheme which may well strike the ignorant with surprise. It was proposed to establish in London a society for ensuring the integrity of clerks, secretaries, collectors, and all such functionaries as are usually obliged to find security for money passing through their hands in the course of business. A gentleman of the highest character as an actuary spoke of the plan in the following terms:—" If a thousand bankers' clerks were to club together to indemnify their securities, by the payment of one pound a year each, and if each had given security for 500*l*., it is obvious that two in each year might become defaulters to that amount, four to half the amount, and so on, without rendering the guarantee fund insolvent. If it be tolerably well ascertained that the instances of dishonesty (yearly) among such persons amount to one in five hundred, this club would continue to exist, subject to being in debt in a bad year, to an amount which it would be able to discharge in good ones. The only question necessary to be asked previous to the formation of such a club would be,—may it not be feared that the motive to resist dishonesty would be lessened by the existence of the club, or that ready-made rogues, by belonging to it, might find the means of obtaining situations which they would otherwise have been kept out of by the impossibility of obtaining security among those who know them? Suppose this be sufficiently answered by saying, that none but those who could bring satisfactory testimony to their previous good character should be allowed to join the club; that persons who may now hope that a deficiency on their parts will be made up and hushed up by the relative or friend who is security, will know very well that the club will have no motive to decline a prosecution, or to keep the secret, and so on. It then only remains to ask, whether the sum demanded for the guarantee is sufficient?" * The philosophical principle on which the scheme proceeds, seems to be simply this, that amongst a given (large) number of persons of good character, there will be, within a year or other considerable space of time, a determinate number of instances in which moral principle and the terror of the consequences of guilt will be overcome by temptations of a determinate kind and

* Dublin Review, Aug. 1840. The Guarantee Society has since been established, and is likely to become a useful and prosperous institution.

amount, and thus occasion a certain periodical amount of loss which the association must make up.

This statistical regularity in moral affairs fully establishes their being under the presidency of law. Man is seen to be an enigma only as an individual; in the mass he is a mathematical problem. It is hardly necessary to say, much less to argue, that mental action, being proved to be under law, passes at once into the category of natural things. Its old metaphysical character vanishes in a moment, and the distinction usually taken between physical and moral is annulled. This view agrees with what all observation teaches, that mental phenomena flow directly from the brain. They are seen to be dependent on naturally constituted and naturally conditioned organs, and thus obedient, like all other organic phenomena, to law. And how wondrous must the constitution of this apparatus be, which gives us consciousness of thought and of affection, which makes us familiar with the numberless things of earth, and enables us to rise in conception and communion to the councils of God himself! It is matter which forms the medium or instrument—a little mass which, decomposed, is but so much common dust; yet in its living constitution, designed, formed, and sustained by

Almighty Wisdom, how admirable its character! how reflective of the unutterable depths of that Power by which it was so formed, and is so sustained!

In the mundane economy, mental action takes its place as a means of providing for the independent existence and the various relations of animals, each species being furnished according to its special necessities and the demands of its various relations. The nervous system—the more comprehensive term for its organic apparatus—is variously developed in different classes and species, and also in different individuals, the volume or mass bearing a general relation to the amount of power. Passing over the humblest orders, where nervous apparatus is so obscure as hardly to be traceable, we see it in the nematoneura of Owen,* in filaments and nuclei, the mere rudiments of the system. In the articulata, it is advanced to a double nervous cord, with ganglia or little masses of nervous matter at frequent intervals, and filaments branching out towards each side; the ganglia near the head being apparently those which send out nerves to the organs of the senses; and this arrangement is only less symmetrical in the mol-

* Including rotifera, entozoa, echinodermata, &c.

lusca. Ascending to the vertebrata, we find a spinal cord, with a brain at the upper extremity, and numerous branching lines of nervous tissue,* an organization strikingly superior; yet here, as in the general structure of animals, the great prin ciple of unity is observed. The brain of the vertebrata is merely an expansion of the anterior pair of the ganglia of the articulata, or these ganglia may be regarded as the rudiment of a brain, the superior organ thus appearing as only a further development of the inferior. There are many facts which tend to prove that the action of this apparatus is of an electric nature, a modification of that surprising agent, which takes magnetism, heat, and light, as other subordinate forms, and of whose general scope in this great system of things we are only beginning to have a faint conception. It has been found that simple electricity, artificially produced, and sent along the nerves of a dead body, excites muscular movement. The brain of a newly-killed animal being taken out, and replaced by a substance which produces

* The ray, which is considered as low in the scale of fishes, and near to the invertebrates, gives the first faint representation of a brain in certain scanty and medullary masses, which appear as merely composed of enlarged origins of the nerves.

electric action, the operation of digestion, which had been interrupted by the death of the animal, was resumed, showing that the brain, in one of its capacities or powers, is identical with the galvanic battery. Nor is this a very startling idea, when we reflect that electricity is almost as metaphysical as ever mind was supposed to be. It is a thing perfectly intangible, weightless. A mass of metal may be magnetized, or heated to seven hundred of Fahrenheit, without becoming the hundredth part of a grain heavier. And yet electricity is a real thing, an actual existence in nature, as witness the effects of heat and light in vegetation—the power of the galvanic current to re-assemble the particles of copper from a solution, and make them again into a solid plate—the rending force of the thunderbolt as it strikes the oak. See also how both heat and light observe the angle of incidence in reflection, as exactly as does a stone thrown obliquely against a wall. So mental action may be imponderable, intangible, and yet a real existence, and ruled by the Eternal through his laws.*

* If mental action is electric, the proverbial quickness of thought—that is, the quickness of the transmission of sensation and will—may be presumed to have been brought to an exact

Common observation shows a great general superiority of the human mind over that of the inferior animals. Man's mind is almost infinite in device ; it ranges over all the world ; it forms the most wonderful combinations ; it seeks back into the past, and stretches forward into the future ; while the animals generally appear to have a narrow range of thought and action. But so also has an infant but a limited range, and yet it is mind which works there, as well as in the most accomplished adults. The difference between mind in the lower animals and in man is a difference in degree only ; it is not a specific difference. All who have studied animals by actual observation, and even those who have given a candid attention to the subject in books, must attain more or less clear convictions of this truth, notwithstanding all

measurement. The speed of light has long been known to be about 192,000 miles per second, and the experiments of Professor Wheatstone have shown that the electric agent travels (if I may so speak) at the same rate, thus showing a likelihood that one law rules the movements of all the "imponderable bodies." Mental action may accordingly be presumed to have a rapidity equal to one hundred and ninety-two thousand miles in the second—a rate evidently far beyond what is necessary to make the design and execution of any of our ordinary muscular movements apparently identical in point of time, which they are.

the obscurity which prejudice may have engendered. We see animals capable of affection, jealousy, envy; we see them quarrel, and conduct quarrels in the very manner pursued by the ruder and less educated of our own race. We see them liable to flattery, inflated with pride, and dejected by shame. We see them as tender to their young as human parents are, and as faithful to a trust as the most conscientious of human servants. The horse is startled by marvellous objects, as a man is. The dog and many others show tenacious memory. The dog also proves himself possessed of imagination, by the act of dreaming. Horses finding themselves in want of a shoe, have of their own accord gone to a farrier's shop where they were shod before. Cats, closed up in rooms, will endeavour to obtain their liberation by pulling a latch or ringing a bell. It has several times been observed that in a field of cattle, when one or two were mischievous, and persisted long in annoying or tyrannizing over the rest, the herd, to all appearance, consulted, and then, making a united effort, drove the troublers off the ground. The members of a rookery have also been observed to take turns in supplying the needs of a family reduced to orphanhood. All of these are acts of reason, in no respect different from similar acts of men. Moreover, although there is no heritage of accumulated knowledge amongst the lower animals, as there is amongst us, they are in some degree susceptible of those modifications of natural character, and capable of those accomplishments, which we call education. The taming and domestication of animals, and the changes thus produced upon their nature in the course of generations, are results identical with civilization amongst ourselves; and the quiet, servile steer is probably as unlike the original wild cattle of this country, as the English gentleman of the present day is unlike the rude baron of the age of King John. Between a young, unbroken horse, and a trained one, there is, again, all the difference which exists between a wild youth reared at his own discretion in the country, and the same person when he has been toned down by long exposure to the influences of refined city society. Of extensive combinations of thought we have no reason to believe that any animals are capable—and yet most of us must feel the force of Walter Scott's remark, that there was scarcely anything which he would not believe of a dog. There is a curious result of education in certain animals, namely, that habits to which they have been trained, in some instances become hereditary. For example, the accomplishment of pointing at game, although a pure result of education, appears in the young pups brought up apart from their parents and kind. The peculiar leap of the Irish horse, acquired in the course of traversing a boggy country, is continued in the progeny brought up in England. This hereditariness of specific habits suggests a relation to that form of psychological manifestation usually called instinct; but instinct is only another term for mind, or is mind in a peculiar stage of development; and though the fact were otherwise, it could not affect the conclusion, that manifestations such as have been enumerated are mainly intellectual manifestations, not to be distinguished as such from those of human beings.

More than this, the lower animals manifested mental phenomena long before man existed. While as yet there was no brain capable of working out a mathematical problem, the economy of the six sided figure was exemplified by the instinct of the bee. The dog and the elephant prefigured the sagacity of the human mind. The love of a human mother for her babe was anticipated by nearly every humbler mammal, the carnaria not excepted. The peacock strutted, the turkey blustered, and the cock fought for victory, just as human beings afterwards did, and still do. Our faculty of imitation, on which so much of our amusement depends, was exercised by the mocking-bird; and the whole tribe of monkeys must have walked about the pre-human world, playing off those tricks in which we see the comicality and mischief-making of our character so curiously exaggerated.

The unity and simplicity which characterize nature give great antecedent probability to what observation seems about to establish, that, as the brain of the vertebrata generally is only an advanced condition of a particular ganglion in the mollusca and crustacea, so are the brains of the higher and more intelligent mammalia only further developments of the brains of the inferior orders of the same class. Or, to the same purpose, it may be said, that each species has certain superior developments, according to its needs, while others are in a rudimental or repressed state. This will more clearly appear after some inquiry has been made into the various powers comprehended under the term mind.

One of the first and simplest functions of mind is to give *consciousness* — consciousness of our identity and of our existence. This, apparently, is independent of the *senses,* which are simply media, and, as Locke has shown, the only media, through which ideas respecting the external world reach the brain. The access of such ideas to the brain is the act to which the metaphysicians have given the name of perception. Gall, however, has shown, by induction from a vast number of actual cases, that there is a part of the brain devoted to perception, and that even this is subdivided into portions which are respectively dedicated to the reception of different sets of ideas, as those of form, size, colour, weight, objects in their totality, events in their progress or occurrence, time, musical sounds, etc. The system of mind invented by this philosopher—the only one founded upon nature, or which even pretends to or admits of that necessary basis—shows a portion of the brain acting as a faculty of comic ideas, another of imitation, another of wonder, one for discriminating or observing differences, and another in which resides the power of tracing effects to causes. There are also parts of the brain for the sentimental part of our nature, or the affections, at the head of which stand the moral feelings of benevolence, conscientiousness, and veneration. Through these, man stands in relation to himself, his fellow-men, the external world, and his God; and through these comes most of the happiness of man's life, as well as that which he derives from the contemplation of the world to come, and the cultivation of his relation to it, (pure religion.) The other sentiments may be briefly enumerated, their names being sufficient in general to denote their functions—firmness, hope, cautiousness, self-esteem, love of approbation, secretiveness, marvellousness, constructiveness, imitation, combativeness, destructiveness, concentrativeness, adhesiveness, love of the opposite sex, love of offspring, alimentiveness, and love of life. Through these faculties, man is connected with the external world, and supplied with active impulses to maintain his place in it as an individual and as a species. There is also a faculty, (language,) for expressing, by whatever means, (signs, gestures, looks, conventional terms in speech,) the ideas which arise in the mind. There is a particular state of each of these faculties, when the ideas of objects once formed by it are revived or reproduced, a process which seems to be intimately allied with some of the phenomena of photography, when images impressed by reflection of the sun's rays upon sensitive paper are, after a temporary obliteration, resuscitated on the sheet being exposed to the fumes of mercury. Such are the phenomena of memory, that handmaid of intellect, without which there could be no accumulation of mental capital, but an universal and continual infancy. Conception and imagination appear to be only intensities, so to speak, of the state of brain in which memory is produced. On their promptness and power depend most of the exertions which distinguish the man of arts and letters, and even in no small measure the cultivator of science.

The faculties above described—the actual elements of the mental constitution—are seen in mature man in an indefinite potentiality and range of action. It is different with the lower animals. They are there comparatively definite in their power and restricted in their application. The reader is familiar with what are called instincts in some of the humbler species, that is, an uniform and unprompted tendency towards certain particular acts, as the building of cells by the bee, the storing of provisions by that insect and several others, and the construction of nests for a coming progeny by birds. This quality is nothing more than a mode of operation peculiar to the faculties in a humble state of endowment, or early stage of development. The cell-formation of the bee, the house-building of ants and beavers, the web-spinning of spiders, are but primitive exercises of constructiveness, the faculty which, indefinite with us, leads to the arts of the weaver, upholsterer, architect, and mechanist, and makes us often work delightedly where our labours are in vain, or nearly so. The storing of provision by the bees is an exercise of acquisitiveness,—a faculty which with us makes rich men and misers. A vast number of curious devices, by which insects provide for the protection and subsistence of their young, whom they are perhaps never to see, are most probably a peculiar restricted effort of philoprogenitiveness. The common source of this class of acts, and of common mental operations, is shown very convincingly by the melting of the one set into the other. Thus, for example, the bee and bird will make modifications in the ordinary form of their cells and nests when necessity compels them. Thus, the alimentiveness of such animals as the dog, usually definite with regard to quantity and quality, can be pampered or educated up to a kind of epicurism, that is, an indefiniteness of object and action. The same faculty acts limitedly in ourselves at first, dictating the special act of sucking; afterwards it acquires indefiniteness. Such is the real nature of the distinction between what are called instinct and reason, upon which so many volumes have been written without profit to the world. All faculties are instinctive, that is, dependent on internal and inherent impulses. This term is therefore not specially applicable to either of the recognised modes of the operation of the faculties. We only, in the one case, see the faculty in an immature and slightly developed state; in the other, in its most advanced condition. In the one case it is *definite*, in the other, *indefinite*, in its range of action. These terms would perhaps be the most suitable for expressing the distinction.

In the humblest forms of being we can trace scarcely anything besides a definite action in a few of the faculties. Generally speaking, as we ascend in the scale, we see more and more of the faculties in exercise, and these tending more to the indefinite mode of manifestation. And for this there is the obvious reason in providence, that the lowest animals have all of them a very limited sphere of existence, born only to perform a few functions, and enjoy a brief term of life, and then give way to another generation, so that they do not need much mental power or guidance. At higher points in the scale, the sphere of existence is considerably extended, and the mental operations are less definite accordingly. The horse, dog, and a few other animals, noted for their serviceableness to our race, have the indefinite powers in no small endowment. Man, again, shows very little of the definite mode of operation, and that little chiefly in childhood, or in barbarism, or idiocy. Destined for a wide field of action, and to be applicable to infinitely varied contingencies, he has all the faculties developed to a high pitch of indefiniteness, that he may be ready to act well in all imaginable cases. His commission, it may be said, gives large discretionary powers, while that of the inferior animals is limited to a few precise directions. But when the human brain is congenitally imperfect or diseased, or when it is in the state of infancy, we see in it an approach towards the character of the brains of some of the inferior animals. Dr. J. G. Davey states that he has frequently witnessed, among his patients at the Hanwell Lunatic Asylum, indications of a particular abnormal cerebration which forcibly reminded him of the specific healthy characteristics of animals lower in the scale of organization;* and every one must have observed how often the actions of children, especially in their moments of play, and where their selfish feelings are concerned, bear a resemblance to those of certain familiar animals.† Behold, then, the wonderful unity of the whole system. The grades of mind, like the forms of being, are mere stages of development. In the humbler forms, but a few of the mental faculties are traceable, just as we see in them but a few of the lineaments of universal structure. In man the system has arrived at its highest condition. The few gleams of reason, then, which we see in the lower animals, are precisely analogous to such a development of the fore-arm as we find in the paddle of the whale. Causality, comparison, and other of the nobler faculties, are in them *rudimental.*

* Phrenological Journal, xv. 338.

† A pampered lap-dog, living where there is another of its own species, will hide any nice morsel which it cannot eat, under a rug, or in some other by-place, designing to enjoy it afterwards. I have seen children do the same thing.

Bound up as we thus are by an identity in the character of our mental organization with the lower animals, we are yet, it will be observed, strikingly distinguished from them by this great advance in development. We have faculties in full force and activity which the animals either possess not at all, or in so low and obscure a form as to be equivalent to non-existence. Now these parts of mind are those which connect us with the things that are not of this world. We have veneration, prompting us to the worship of the Deity, which the animals lack. We have hope, to carry us on in thought beyond the bounds of time. We have reason, to enable us to inquire into the character of the Great Father, and the relation of us, his humble creatures, towards him. We have conscientiousness and benevolence, by which we can in a faint and humble measure imitate, in our conduct, that which he exemplifies in the whole of his wondrous doings. Beyond this, mental science does not carry us in support of religion: the rest depends on evidence of a different kind. But it is surely much that we thus discover in nature a provision for things so important. The existence of faculties having a regard to such things is a good evidence that such things exist. The face of God is reflected in the organization of man, as a little pool reflects the glorious sun.

The affective or sentimental faculties are all of them liable to operate whenever appropriate objects or stimuli are presented, and this they do as irresistibly and unerringly as the tree sucks up moisture which it requires, with only this exception, that one faculty often interferes with the action of another, and operates instead, by force of superior inherent strength or temporary activity. For example, alimentiveness may be in powerful operation with regard to its appropriate object, producing a keen appetite, and yet it may not act, in consequence of the more powerful operation of cautiousness, warning against evil consequences likely to ensue from the desired indulgence. This liability to flit from under the control of one feeling to the control of another, constitutes what is recognised as free will in man, being nothing more than a vicissitude in the supremacy of the faculties over each other.

It is a common mistake to suppose that the individuals of our own species are all of them formed with similar faculties—similar in power and tendency—and that education and the influence of circumstances produce all the differences which we observe. There is not, in the old systems of mental philosophy, any doctrine more opposite to the truth than this. It is refuted at once by the great differences of intellectual tendency and moral disposition to be observed amongst a group of young children who have been all brought up in circumstances perfectly identical — even in twins, who have never been but in one place, under the charge of one nurse, attended to alike in all respects. The mental characters of individuals are inherently various, as the forms of their persons and the features of their faces are; and education and circumstances, though their influence is not to be despised, are incapable of entirely altering these characters, where they are strongly developed. That the original characters of mind are dependent on the volume of particular parts of the brain and the general quality of that viscus, is proved by induction from an extensive range of observations, the force of which must have been long since universally acknowledged but for the unpreparedness of mankind to admit a functional connexion between mind and body. The different mental characters of individuals may be presumed from analogy to depend on the same law of development which we have seen determining forms of being and the mental characters of particular species. This we may conceive as carrying forward the intellectual powers and moral dispositions of some to a high pitch, repressing those of others at a moderate amount, and thus producing all the varieties which we see in our fellow-creatures. Thus a Cuvier and a Newton are but expansions of a clown, and the person emphatically called the wicked man, is one whose highest moral feelings are rudimental. Such differences are not confined to our species; they are only less strongly marked in many of the inferior animals. There are clever dogs and wicked horses, as well as clever men and wicked men; and education sharpens the talents, and in some degree regulates the dispositions of animals, as it does our own.

There is, nevertheless, a general adaptation of the mental constitution of man to the circumstances in which he lives, as there is between all the parts of nature to each other. The goods of the physical world are only to be realized by ingenuity and industrious exertion; behold, accordingly, an intellect full of device, and a fabric of the faculties which would go to pieces or destroy itself if it were not kept in constant occupation. Nature presents to us much that is sublime and beautiful: behold faculties which delight in contemplating these properties of hers, and in rising upon them, as upon wings, to the presence of the Eternal. It is also a world of difficulties and perils, and see how a large portion of our species are endowed with vigorous powers, which take a pleasure in meeting and overcoming difficulty and danger. Even that principle on which our faculties are constituted—a wide range of freedom in which to act for all various occasions—necessitates a resentful faculty, by which individuals may protect themselves from the undue and capricious exercise of each other's faculties, and thus preserve their individual rights. So also there is cautiousness, to give us a tendency to provide against the evils by which we may be assailed; and secretiveness, to enable us to conceal whatever, being divulged, would be offensive to others or injurious to ourselves,—a function which obviously has a certain legitimate range of action, however liable to be abused. The constitution of the mind generally points to a state of intimate relation of individuals towards society, towards the external world, and towards things above this world. No individual being is integral or independent; he is only part of an extensive piece of social mechanism. The inferior mind, full of rude energy and unregulated impulse, does not more require a superior nature to act as its master and its mentor, than does the superior nature require to be surrounded by such rough elements on which to exercise its high endowments as a ruling and tutelary power. This relation of each to each produces a vast portion of the active business of life. It is easy to see that, if we were all alike in our moral tendencies, and all placed on a medium of perfect moderation in this respect, the world would be a scene of everlasting dullness and apathy. It requires the variety of individual constitution to give moral life to the scene.

The indefiniteness of the potentiality of the human faculties, and the complexity which thus attends their relations, lead unavoidably to occasional error. If we consider for a moment that there are not less than thirty such faculties, that they are each given in different proportions to different persons, that each is at the same time endowed with a wide discretion as to the force and frequency of its action, and that our neighbours, the world, and our connexions with something beyond it, are all exercising an ever-varying influence over us, we cannot be surprised at the irregularities attending human conduct. It is simply the penalty paid for the superior endowment. It is here that the so-called imperfection of our nature resides. Causality and conscientiousness are, it is true, guides over all; but even these are only faculties of the same indefinite potentiality as the rest, and partake accordingly of the same inequality of action. Man is therefore a piece of mechanism, which never can act so as to satisfy his own ideas of what he might be—for he can imagine a state of moral perfection, (as he can imagine a globe formed of diamonds, pearls, and rubies,) though his constitution forbids him to realize it. There ever will, in the best disposed and most disciplined minds, be occasional discrepancies between the amount of temptation and the power summoned for regulation or resistance, or between the stimulus and the mobility of the faculty; and hence those errors, and shortcomings, and excesses, without end, with which the good are constantly finding cause to charge themselves. There is at the same time even here a possibility of improvement. In infancy, the impulses are all of them irregular; a child is cruel, cunning, and false, under the slightest temptation, but in time learns to control these inclinations, and to be habitually humane, frank, and truthful. So is human society, in its earliest stages, sanguinary, aggressive, and deceitful, but in time, becomes just, faithful, and benevolent. To such improvements there is a natural tendency which will operate in all fair circumstances, though it is not to be expected that irregular and undue impulses will ever be altogether banished from the system.

It may still be a puzzle to many, how beings should be born into the world whose organization is such that they unavoidably, even in a civilized country, become malefactors. Does God, it may be asked, make criminals? Does he fashion certain beings with a predestination to evil? He does not do so; and yet the criminal type of brain, as it is called, comes into existence in accordance with laws which the Deity has established. It is not, however, as the result of the first or general intention of those laws, but as an exception from their ordinary and proper action. The production of those evilly disposed beings is in this manner. The moral character of the progeny depends in a general way, (as does the physical character also,) upon conditions of the parents,—both general conditions, and conditions at the particular time of the commencement of the existence of the new being, and likewise external conditions affecting the fœtus through the mother. Now the amount of these conditions is indefinite. The faculties of the parents, as far as these are concerned, may have oscillated for the time towards the extreme of tensibility in one direction. The influences upon the fœtus may have also been of an extreme and unusual kind. Let us suppose that the conditions upon the whole have been favourable for the development, not of the higher, but of the lower sentiments, and of the propensities of the new being, the result will necessarily be a mean type of brain. Here, it will be observed, God no more decreed an immoral being, than he decreed an immoral paroxysm of the sentiments. Our perplexity is in considering the ill-disposed being by himself. He is only a part of a series of phenomena, traceable to a principle good in the main, but which admits of evil as an exception. We have seen that it is for wise ends that God leaves our moral faculties to an indefinite range of action: the general good results of this arrangement are obvious; but exceptions of evil are inseparable from such a system, and this is one of them. To come to particular illustration—when a people are oppressed, or kept in a state of slavery, they invariably contract habits of lying, for the purpose of deceiving and outwitting their superiors, falsehood being a refuge of the weak under difficulties. What is a habit in parents becomes an inherent quality in children. We are not, therefore, to be surprised when a traveller tells us that black children in the West Indies appear to lie by instinct, and never answer a white person truly, even in the simplest matter. Here we have secretiveness roused in a people to a state of constant and exalted exercise; an over tendency of the nervous energy in that direction is the consequence, and a new organic condition is established. This tells upon the progeny, which comes into the world with secretiveness excessive in strength and activity. All other evil characteristics may be readily conceived as being implanted in a new generation in the same way. And sometimes not one, but several generations, may be concerned in bringing up the result to a pitch which produces crime. It is, however, to be observed, that the general tendency of things is to a limitation, not the extension of such abnormally constituted beings. The criminal brain finds itself in a social scene where all is against it. It may struggle on for a time, but it is sure to be overcome at last by the medium and superior natures. The disposal of such beings will always depend much on the moral state of a community, the degree in which just views prevail with regard to human nature, and the feelings which accident may have caused to predominate at a particular time. Where the mass was little enlightened or refined, and terrors for life or property were highly excited, malefactors have ever been treated severely. But when order is generally triumphant, and reason allowed sway, men begin to see the true case of criminals—namely, that while one large section are victims of erroneous social conditions, another are brought to error by tendencies which they are only unfortunate in having inherited from nature. Criminal jurisprudence then addresses itself less to the direct punishment than to the reformation and care-taking of those liable to its attention. And such a treatment of criminals, it may be farther remarked, so that it stop short of affording any encouragement to crime, (a point which experience will determine,) is evidently no more than justice, seeing how accidentally all forms of the moral constitution are distributed, and how thoroughly mutual obligation shines throughout the whole frame of society—the strong to help the weak, the good to redeem and restrain the bad.

The sum of all we have seen of the psychical constitution of man is, that its Almighty Author has destined it, like everything else, to be developed from inherent qualities, and to have a mode of action depending solely on its own organization. Thus the whole is complete on one principle. The masses of space are formed by law; law makes them in due time theatres of existence for plants and animals; sensation, disposition, intellect, are all in like manner developed and sustained in action by law. It is most interesting to observe into how small a field the whole of the mysteries of nature thus ultimately resolve themselves. The inorganic has been thought to have one final comprehensive law, GRAVITATION. The organic, the other great department of mundane things, rests in like manner on one law, and that is—DEVELOPMENT. Nor may even these be after all twain, but only branches of one still more comprehensive law, the expression of a unity, flowing immediately from the One who is First and Last.

PURPOSE AND GENERAL CONDITION OF THE ANIMATED CREATION.

We have now to inquire how this view of the constitution and origin of nature bears upon the condition of man upon the earth, and his relation to supra-mundane things.

That enjoyment is the proper attendant of animal existence is pressed upon us by all that we see and all we experience. Everywhere we perceive in the lower creatures, in their ordinary condition, symptoms of enjoyment. Their whole being is a system of needs, the supplying of which is gratification, and of faculties, the exercise of which is pleasurable. When we consult our own sensations, we find that, even in a sense of a healthy performance of all the functions of the animal economy, God has furnished us with an innocent and very high enjoyment. The mere quiet consciousness of a healthy play of the mental functions—a mind at ease with itself and all around it—is in like manner extremely agreeable. This negative class of enjoyments, it may be remarked, is likely to be even more extensively experienced by the lower animals than by man, at least in the proportion of their absolute endowments, as their mental and bodily functions are much less liable to derangement than ours. To find the world constituted on this principle is only what in reason we should expect. We cannot conceive that so vast a system could have been created for a contrary purpose. No averagely constituted human being would, in his own limited sphere of action, think of producing a similar system upon an opposite principle. But to form so vast a range of being, and to make being everywhere a source of gratification, is conformable to our ideas of a Creator, in whom we are constantly discovering traits of a nature, of which our own is a faint and far-cast shadow.

It appears at first difficult to reconcile with this idea the many miseries which we see all sentient beings, ourselves included, occasionally enduring. How, the sage has asked in every age, should a

Being so transcendently kind, have allowed of so large an admixture of evil in the condition of his creatures? Do we not at length find an answer to a certain extent satisfactory, in the view which has now been given of the constitution of nature? We there see the Deity operating in the most august of his works by fixed laws, an arrangement which, it is clear, only admits of the main and primary results being good, but disregards exceptions. Now the mechanical laws are so definite in their purposes, that no exceptions ever take place in that department; if there is a certain quantity of nebulous matter to be agglomerated and divided and set in motion as a planetary system, it will be so with hair's-breadth accuracy, and cannot be otherwise. But the laws presiding over meteorology, life, and mind, are necessarily less definite, as they have to produce a great variety of mutually related results. Left to act independently of each other, each according to its separate commission, and each with a wide range of potentiality to be modified by associated conditions, they can only have effects generally beneficial. Often there must be an interference of one law with another; often a law will chance to operate in excess, or upon a wrong object, and thus evil will be produced. Thus, winds are generally useful in many ways, and the sea is useful as a means of communication between one country and another; but the natural laws which produce winds are of indefinite range of action, and sometimes are unusually concentrated in space or in time, so as to produce storms and hurricanes, by which much damage is done; the sea may be by these causes violently agitated, so that many barks and many lives perish. Here, it is evident, the evil is only exceptive. Suppose, again, that a boy, in the course of the lively sports proper to his age, suffers a fall which injures his spine, and renders him a cripple for life. Two things have been concerned in the case: first, the love of violent exercise, and second, the law of gravitation. Both of these things are good in the main. Boys, in the rash enterprises and rough sports in which they engage, are only making the first delightful trials of a bodily and mental energy which has been bestowed upon them as necessary for their figuring properly in a scene where many energies are called for, and where the exertion of these powers is ever a source of happiness. By gravitation, all moveable things, our own bodies included, are kept stable on the surface of the earth. But when it chances that the playful boy loses his hold (we shall say) of the branch of a tree, and has no solid support immediately below, the law of gravitation unrelentingly pulls him to the ground, and thus he is hurt. Now it was not a primary object of gravitation to injure boys; but gravitation could not but operate in the circumstances, its nature being to be universal and invariable. The evil is, therefore, only a casual exception from something in the main good.

The same explanation applies to even the most conspicuous of the evils which afflict society. War, it may be said, and said truly, is a tremendous example of evil, in the misery, hardship, waste of human life, and mis-spending of human energies, which it occasions. But what is it that produces war? Certain tendencies of human nature, as keen assertion of a supposed right, resentment of supposed injury, acquisitiveness, desire of admiration, combativeness, or mere love of excitement. All of these are tendencies which are every day, in a legitimate extent of action, producing great and indispensable benefits to us. Man would be a tame, indolent, unserviceable being without them, and his fate would be starvation. War, then, huge evil though it be, is, after all, but the exceptive case, a casual misdirection of properties and powers essentially good. God has given us the tendencies for a benevolent purpose. He has only not laid down any absolute obstruction to our misuse of them. That were an arrangement of a kind which he has nowhere made. But he has established many laws in our nature which tend to lessen the frequency and destructiveness of these abuses. Our reason comes to see that war is purely an evil, even to the conqueror. Benevolence interposes to make its ravages less mischievous to human comfort, and less destructive to human life. Men begin to find that their more active powers can be exercised with equal gratification on legitimate objects ; for example, in overcoming the natural difficulties of their path through life, or in a generous spirit of emulation in a line of duty beneficial to themselves and their fellow-creatures. Thus, war at length shrinks into a comparatively narrow compass, though there certainly is no reason to suppose that it will be at any early period, if ever, altogether dispensed with, while man's constitution remains as it is. In considering an evil of this kind, we must not limit our view to our own or any past time. Placed upon the earth with faculties prepared to act, but inexperienced, and with the more active propensities necessarily in great force to suit the condition of the globe, man was apt to misuse his powers much in this way at first, compared with what he is likely to do when he advances into a condition of civilization. In the scheme of providence, thousands of years of frequent warfare, all the so-called glories which fill history, may be but a subordinate consideration. The chronology of God is not as our chronology. See the patience of waiting evinced in the slow development of the animated kingdoms, throughout the long series of geological ages. Nothing is it to him that an entire goodly planet should, for an inconceivable period, have no inhabiting organisms superior to reptiles. Nothing is it to him that whole astral systems should be for infinitely longer spaces of time in the nebular embryo, unfit for the reception of one breathing or sentient being out of the myriad multitudes who are yet to manifest his goodness and his greatness. Progressive, not instant effect is his sublime rule. What, then, can it be to him that the human race goes through a career of impulsive acting for a few thousand years? The cruelties of ungoverned anger, the tyrannies of the rude and proud over the humble and good, the martyr's pains, and the patriot's despair, what are all these but incidents of an evolution of superior being which has been pre-arranged and set forward in independent action, free within a certain limit, but in the main constrained, through primordial law, to go on ever brightening and perfecting, yet never, while the present dispensation of nature shall last, to be quite perfect!

The sex passion in like manner leads to great evils. Providence has seen it necessary to make very ample provision for the preservation and utmost possible extension of all species. The aim seems to be to diffuse existence as widely as possible, to fill up every vacant piece of space with some sentient being to be a vehicle of enjoyment. Hence this passion is conferred in great force. But the relation between the number of beings, and the means of supporting them, is only on the footing of general law. There may be occasional discrepancies between the laws operating for the multiplication of individuals, and the laws operating to supply them with the means of subsistence, and evils will be endured in consequence, even in our own highly favoured species. But against all these evils, and against those numberless vexations which have arisen in all ages from the attachment of the sexes, place the vast amount of happiness which is derived from this source—the basis of the whole circle of the domestic affections, the sweetening principle of life, the prompter of all our most generous feelings, and even of our most virtuous resolves and exertions—and every ill that can be traced to it is but as dust in the balance. And here, also, we must be on our guard against judging from what we see in the world at a particular era. As reason and the higher sentiments of man's nature increase in force, this passion is put under better regulation, so as to lessen many of the evils connected with it. The civilized man is more able to give it due control; his attachments are less the result of impulse: he studies more the weal of his partner and offspring. There are even some of the resentful feelings connected in early society with love, such as hatred of successful rivalry, and jealousy, which almost disappear in an advanced state of civilization. The evils springing, in our own species at least, from this passion, may therefore be an exception mainly peculiar to a particular term of the world's progress, and which may be expected to decrease greatly in amount.

With respect, again, to disease, so prolific a cause of suffering to man, the human constitution is merely a complicated but regular process in electro-chemistry, which goes on well, and is a source of continual gratification, so long as nothing occurs to interfere with it injuriously, but which is liable every moment to be deranged by various external agencies, when it becomes a source of pain, and, if the injury be severe, ceases to be capable of retaining life. It may be readily admitted that the evils experienced in this way are very great; but, after all, such experiences are no more than occasional, and not necessarily frequent —exceptions from a general rule of which the direct action is to confer happiness. The human constitution might have been made of a more hardy character; but we always see hardiness and insensibility go together, and it may be of course presumed that we only could have purchased this immunity from suffering at the expense of a large portion of that delicacy in which lie some of our most agreeable sensations. Or man's faculties might have been restricted to definiteness of action, as is greatly the case with those of the lower animals, and thus we should have been equally safe from the aberrations which lead to disease; but in that event we should have been incapable of acting to so many different purposes as we are, and of the many high enjoyments which the varied action of our faculties places in our power; we should not, in short, have been human beings, but merely on a level with the inferior animals. Thus, it appears, that the very fineness of man's constitution, that which places him in such a high relation to the mundane economy, and makes him the vehicle of so many exquisitely delightful sensations—it is this which makes him liable to the sufferings of disease. It might be said, on the other hand, that the noxiousness of the agencies producing disease might have been diminished or extinguished; but the probability is, that this could not have been done without such a derangement of the whole economy of nature as would have been attended with more serious evils. For example—a large class of diseases are the result of effluvia from decaying organic matter. This kind of matter is known to be extremely useful when mixed with earth, in favouring the process of vegetation. Supposing the noxiousness to the human constitution done away with, might we not also lose that important quality which tends so largely to increase the food raised from the ground? Perhaps (as has been somewhere suggested) the noxiousness is even a matter of special design, to induce us to put away decaying organic substances into the earth, where they are calculated to be so useful. Now man has reason to enable him to see that such substances are beneficial under one arrangement, and noxious in the other. He is, as it were, commanded to take the right method in dealing with them. In point of fact, men do not always take this method, but allow accumulations of noxious matter to gather close about their dwellings, where they generate fevers and agues. But their doing so may be regarded as only a temporary exception from the operation of mental laws, the general tendency of which is to make men adopt the proper measures. And these measures will probably be in time universally adopted, so that one extensive class of diseases will be altogether or nearly abolished.

Another large class of diseases spring from mismanagement of our personal economy. Eating to excess, eating and drinking what is noxious, disregard to that cleanliness which is necessary for the right action of the functions of the skin, want of fresh air for the supply of the lungs, undue, excessive, and irregular indulgence of the mental affections, are all of them recognised modes of creating that derangement of the system in which disease consists. Here also it may be said that a limitation of the mental faculties to definite manifestations (*vulgo*, instincts) might have enabled us to avoid many of these errors; but here again we are met by the consideration that, if we had been so endowed, we should have been only as the lower animals are,—wanting that transcendently higher character of sensation and power, by which our enjoyments are made so much greater. In making the desire of food, for example, with us an indefinite mental manifestation, instead of the definite one, which it mainly is amongst the lower animals, the Creator has given us a means of deriving far greater gratifications from food (consistently with health) than the lower animals generally appear to be capable of. He has also given us reason to act as a guiding and controlling power over this and other propensities, so that they may be prevented from becoming causes of malady. We can see that excess is injurious, and are thus prompted to moderation. We can see that all the things which we feel inclined to take are not healthful, and are thus exhorted to avoid what are pernicious. We can also see that a cleanly skin and a constant supply of pure air are necessary to the proper performance of some of the most important of the organic functions, and thus are stimulated to frequent ablution, and to a right ventilation of our parlours and sleeping apartments. And so on with the other causes of disease. Reason may not operate very powerfully to these purposes in an early state of society, and prodigious evils may therefore have been endured from diseases in past ages ; but these are not necessarily to be endured always. As civilization advances, reason acquires a greater ascendancy; the causes of the evils are seen and avoided ; and disease shrinks into a comparatively narrow compass. The experience of our own country places this in a striking light. In the middle ages, when large towns had no police regulations, society was at frequent intervals scourged by pestilence. The third part of the people of Europe are said to have been carried off by one epidemic. Even in London the annual mortality has greatly sunk within a century. The improvement in human life, which has taken place since the construction of the Northampton tables by Dr. Price, is equally remarkable. Modern tables still show a prodigious mortality among the young in all civilized countries—evidently a result of some prevalent error in the usual modes of rearing them. But to remedy this evil there is the sagacity of the human mind, and the sense to adopt any reformed plans which may be shown to be necessary. By a change in the management of an orphan institution in London, during the last fifty years, an immense reduction in the mortality took place. We may of course hope to see measures devised and adopted for producing a similar improvement of infant life throughout the world at large.

In this part of our subject, the most difficult point certainly lies in those occurrences of disease where the afflicted individual has been in no degree concerned in bringing the visitation upon himself. Daily experience shows us infectious disease arising in a place where the natural laws in respect of cleanliness are neglected, and then spreading into regions where there is no blame of this kind. We then see the innocent suffering equally with those who may be called the guilty. Nay, the benevolent physician who comes to succour the miserable beings whose error may have caused the mischief, is sometimes seen to fall a victim to it, while many of his patients recover. We are also only too familiar with the transmission of diseases from erring parents to innocent children, who accordingly suffer, and perhaps die prematurely, as it were for the sins of others. After all, however painful such cases may be in contemplation, they cannot be regarded in any other light than as exceptions from arrangements, the general working of which is beneficial.

With regard to the innocence of the suffering parties, there is one important consideration which is pressed upon us from many quarters—namely, that moral conditions have not the least concern in the working of the physical laws. These arrangements proceed with an entire independence of all such conditions, and desirably so, for otherwise there could be no certain dependence placed upon them. Thus it may happen that two persons ascending a piece of scaffolding, the one a virtuous, the other a vicious man, the former, being the less cautious of the two, ventures upon an insecure place, falls, and is killed, while the other, choosing a better footing, remains uninjured. It is not in what we can conceive of the nature of things, that there should be a special exemption from the ordinary laws of matter, to save this virtuous man. So it might be that, of two physicians, attending fever cases, in a mean part of a large city, the one, an excellent citizen, may stand in such a position with respect to the beds of the patients as to catch the infection, of which he dies in a few days, while the other, a bad husband and father, and who, unlike the other, only attends such cases with selfish ends, takes care to be as much as possible out of the stream of infection, and accordingly escapes. In both of these cases man's sense of good and evil—his faculty of conscientiousness—would incline him to destine the vicious man to destruction and save the virtuous. But the Great Ruler of Nature does not act on such principles. He has established laws for the operation of inanimate matter, which are quite unswerving, so that, when we know them, we have only to act in a certain way with respect to them, in order to obtain all the benefits and avoid all the evils connected with them. He has likewise established moral laws in our nature, which are equally unswerving, (allowing for their wider range of action,) and from obedience to which unfailing good is to be derived. But the two sets of laws are independent of each other. Obedience to each gives only its own proper advantage, not the advantage proper to the other. Hence it is that virtue forms no protection against the evils connected with the physical laws, while, on the other hand, a man skilled in, and attentive to these, but unrighteous and disregardful of his neighbour, is in like manner not protected by his attention to physical circumstances from the proper consequences of neglect or breach of the moral laws.

Thus it is that the innocence of the party suffering for the faults of a parent, or of any other person or set of persons, is evidently a consideration quite apart from that suffering.

In short, the whole question of evil, a puzzle throughout all ages, only becomes explicable when we receive and study the system of a mundane government by law. There is no need for considering it as a detraction from either the power or the goodness of God. The dispensation under which we live has been constituted by him on the principle of law; but this is not necessarily to imply that either his goodness or his power is to stop at this point. That such, however, is the character of the pageantry of worldly events now passing, is the only idea we can arrive at when we approach the question without prejudice. How else should it be that in any case the guilty flourish and the innocent suffer? How else should it be that men often endure bitter woe and pain while prosecuting the noblest objects? How else should we ever see so simple an event as the following, which meets my eyes in the journals, while these sheets pass through the press:—A multitude of poor Irish emigrants are embarked in a canal boat, about to leave their native district for a port whence they are to sail for America. At the moment of parting, they crowd to one side, to shake hands for the last time with their friends. The vessel is overbalanced and turned upon its side. Of the multitude thrown into the water, seven are taken up dead. Here, an action rather amiable and laudable than otherwise, leads to the loss of life,—a pure evil, unmixed with good. It is impossible to imagine such a transaction occurring under the immediate direction of the Deity; it would be profaning human nature to attribute any such act to the immediate command or interference of a man. But there is no difficulty in understanding how such occasional evils should take place in the course of a chain of causes which only proceed in consequence of a remote and general impulse designed in the main for good.

Evil, indeed, is one of the strongest proofs that could be desired for the reality of this system. We see it in one of its most familiar forms in the destructive animals. An innocent little bird in the claws of the cruel hawk—a poor stag grasped by the ruthless boa—a lamb in the fangs of the wolf—can we imagine a form of misery greater than the fate of these animals? Yet millions of such creatures perish in this manner annually, and have so done since long before there existed a human heart to pine or break with its more sentimental, but not less real wretchedness. Upon no theory can this be understood except upon that of an economy governed by general laws. The carnivorous animals are simply the police and undertakers of the inferior creation, preventing their too great increase, and clearing off all such as grow weakly and die, ere they can become in any degree a burden to themselves or a nuisance to other creatures. For these functions the destructive tribes have been expressly organized, and their organization of course is of divine appointment. Constituted as we are, we cannot suppose a plan involving so much suffering to have been adopted except with a view to that independency, or completeness within itself, which is here argued for as the manner in which the Deity's operations on earth are revealed to us. He has endowed the families which enjoy his bounties with an almost indefinite fecundity, that enjoyment may be as widely diffused as possible;

but the limitation of the results of this fecundity within the line necessary according to circumstances, were no right immediate employment for himself. The object is accomplished, in a befitting manner, by his ordaining that certain other animals shall have endowments sure so to act as to bring the rest of animated beings to a proper balance. And the object is accomplished well; insomuch that we never hear of any but the most partial and transient discrepancy between the volume of inferior animal life and the power appointed for its regulation. Even in this painful chapter of nature, we are forced to acknowledge that, upon the theory of law, every thing is very good.

Another proof, or rather another branch of the same proof, lies in the relation of the individual to the mass, as far as endowment and destiny are concerned. We see, for example, powerful impulses in human nature, which often occasion great inconveniences both to those yielding to them and to others. But such impulses are in the main necessary. Destructive, in many cases, to the individual, they are conservative with respect to the totality. What is this but an appointment to render the machine in that respect a self-acting one? Many of the confusions of the moral scene might be thus explained; but it is also to be observed that such impulses are not sent alone—they come in company with intelligence and moral emotion, powers continually tending more and more to soften and regulate their action.

Nor are any of the ordinary evils of our world altogether unmixed. God, contemplating apparently the unbending action of his great laws, has established others which appear to be designed to have a compensating, a repairing, and a consoling effect. Suppose, for instance, that, from a defect in the power of development in a mother, her offspring is ushered into the world destitute of some of the most useful members, or blind, or deaf, or of imperfect intellect, there is ever to be found in the parents and other relatives, and in the surrounding public, a sympathy with the sufferer, which tends to make up for the deficiency, so that he usually is in the long run not much a loser. Indeed, the benevolence implanted in our nature seems to be an arrangement having for one of its principal objects to cause us, by sympathy and active aid, to remedy the evils unavoidably suffered by our fellow-creatures in the course of the operation of the other natural laws. And even in the sufferer himself, it is often found that a defect in one point is made up for by an extra power in another. The blind come to have a sense of touch much more acute than those who see. Persons born without hands have been known to acquire a power of using their feet for a number of the principal offices usually served by that member. I need hardly say how remarkably fatuity is compensated by the more than usual regard paid to the children born with it by their parents, and the zeal which others usually feel to protect and succour such persons. In short, we never see evil of any kind take place where there is not some remedy or compensating principle ready to interfere for its alleviation. And there can be no doubt that in this manner suffering of all kinds is very much relieved.

We may, then, regard the globes of space as theatres designed for the residence of animated sentient beings, placed there with this as their first and most obvious purpose—to be sensible of enjoyments from the exercise of their faculties in relation to external things. The faculties of the various species are very different, but the happiness of each depends on the harmony there may be between its particular faculties and its particular circumstances. For instance, place the small-brained sheep or ox in a good pasture, and it fully enjoys this harmony of relation; but man, having many more faculties, cannot be thus contented. Besides having a sufficiency of food and bodily comfort, he must have entertainment for his intellect, whatever be its grade, objects for the domestic and social affections, objects for the sentiments. He is also a progressive being, and what pleases him to-day may not please him to-morrow; but, in each case he demands a sphere of appropriate conditions in order to be happy. By virtue of his superior organization, his enjoyments are much higher and more varied than those of any of the lower animals; but the very complexity of circumstances affecting him renders it at the same time unavoidable, that his nature should be often inharmoniously placed and disagreebly affected, and that he should therefore be unhappy. Still, unhappiness amongst mankind is the exception from the rule of their condition, and an exception which is capable of almost infinite diminution, by virtue of the improving reason of man, and the experience which he acquires in working out the problems of society.

To secure the immediate means of happiness, it would seem to be necessary for men first to study with all care the constitution of nature; and, secondly, to accommodate themselves to that constitution, so as to obtain all the realizable advantages from acting conformably to it, and to avoid all likely evils from disregarding it. It will be of no use to sit down and expect that things are to operate of their own accord, or through the direction of a partial deity, for our benefit; equally so were it to expose ourselves to palpable dangers, under the notion that we shall, for some reason, have a dispensation or exemption from them: we must endeavour so to place ourselves, and so to act, that the arrangements which Providence has made impartially for all may be in our favour, and not against us; such are the only means by which we can obtain good and avoid evil here below.* And, in doing this, it is especially necessary that care be taken to avoid interfering with the like efforts of other men, beyond what may have been

* The doctrine of the natural laws as affecting human welfare is clearly and satisfactorily explained in Mr. Combe's *Essay on the Constitution of Man*, to which and to the excellent works of Dr. Andrew Combe, may be ascribed no small share of that public movement towards improved sanitary regulations which is one of the most cheering features of our age.

agreed upon by the mass as necessary for the general good. Such interferences, tending in any way to injure the body, property, or peace of a neighbour, or to the injury of society in general, tend to reflect evil upon ourselves through the re-action which they produce in the feelings of our neighbour and of society, and also the offence which they give to our own conscientiousness and benevolence. On the other hand, when we endeavour to promote the efforts of our fellow-creatures to attain happiness, we produce a re-action of the contrary kind, the tendency of which is towards our own benefit. The one course of action tends to the injury, the other to the benefit of ourselves and others. By the one course, the general design of the Creator towards his creatures is thwarted; by the other it is favoured. And thus we can readily see the most substantial grounds for regarding all moral emotions and doings as divine in their nature, and as a means of rising to and communing with God. Obedience is not selfishness, which it would otherwise be—it is worship. The merest barbarians have a glimmering sense of this philosophy, and it continually shines out more and more clearly as men advance in intelligence. Nor are individuals alone concerned here. The same rule applies as between one great body or class of men and another, and also between nations. Thus, if one set of men keep others in the condition of slaves—this being a gross injustice to the subjected party, the mental manifestations of that party to the masters will be such as to mar the comfort of their lives; the minds of the masters themselves will be degraded by the association with beings so degraded; and thus, with some immediate or apparent benefit from keeping slaves, there will be in a far greater degree an experience of evil. So also, if one portion of a nation, engaged in a particular department of industry, grasp at some advantages injurious to the other sections of the people, the first effect will be an injury to those other portions of the nation, and the second a re-active injury to the injurers, making their guilt their punishment. And so when one nation commits an aggression upon the property or rights of another, or even pursues towards it a sordid or ungracious policy, the effects are sure to be redoubled evil from the offended party. All of these things are under laws which make the effects, on a large range, absolutely certain; and an individual, a party, a people, can no more act unjustly with safety, than I could with safety place my leg in the track of a coming wain, or attempt to fast thirty days. We have been constituted on the principle of only being able to realize happiness for ourselves when our fellow-creatures are also happy; it is therefore necessary that we both do to others only as we would have others to do to us, and endeavour to promote their happiness as well as our own. There is even a higher law, which has long been announced, but never acted on to any considerable extent, that our greatest happiness is not to be realized by each having a regard for himself, but by each seeking primarily to benefit his fellow-creatures. When man comes to have confidence in his own nature, he will begin to act on this principle, and the result will be a degree of happiness such as we only see at present faintly shadowed forth in the purest and sweetest charities of life—a happiness from which there will be no class exceptions.

The question whether the human race will ever advance far beyond its present position in intellect and morals, is one which has engaged much attention. Judging from the past, we cannot reasonably doubt that great advances are yet to be made; but if the principle of development be admitted, these are certain, whatever may be the space of time required for their realization. A progression resembling development may be traced in human nature, both in the individual and in large groups of men. The individual is in childhood under the influence of the propensities and instinctive aptitudes; in youth, he is swayed by marvellousness, the love of the beautiful, the imaginative; in full maturity, he passes under (comparatively) the domination of reason. In perfect analogy, a nation is at first impulsive and unreasoning; afterwards it is conducted by the second class of sentiments, (the age of mythologies, hierocracies, man and idea worships;) finally, its institutions approximate to an accurate regard for what is convenient and profitable, under the control of justice and humanity. The advance of knowledge favours the progress of the moral conditions, and in improved moral conditions knowledge becomes more sound. In tolerably favourable circumstances, this tendency onward never fails to make itself visible; and it is evident that, though many nations seem nearly stationary and others appear to retrograde, there is always a progress in some place, so that no long space of time ever elapses without showing, upon the whole, a certain advance. Now all this is in conformity with what we have seen of the progress of organic creation. It seems but the minute hand of a watch, of which the hour hand is the transition from species to species. Knowing what we do of that latter transition, the possibility of a decided and general retrogression of the highest species towards a meaner type is scarce admissible, but a forward movement seems anything but unlikely. This view is favoured even by zoological science. We there see order after order of animals, from the bottom of the scale upwards, consisting of many genera, each of these again presenting various species, until we come to the highest order of all—BIMANA; and behold of this order but one genus—nay, but one species to represent that genus, namely, Man! Take any of the highest orders next to man—the Lemuridæ, the Vespertilionidæ, the Simiadæ, and into what multitudes of species do we find them varying! The Bimana alone is *of one species.* For this no shadow of a zoological reason can be presented. It is supported by none of the analogies of nature, but, on the contrary, is in decided contradiction to them, that there should be but one species of the highest type of animated being. The zoological series appears here, as it were, broken short, or interrupted in its progress towards a general symmetry. Is not this a strong indication of further progress in development being designed? Is not the right explanation simply this—that the animated creation is seen by us at *a particular point in its progress?*—a progress yet to be continued. To this conclusion, all our knowledge of the past external conditions of the earth conduces. We there see ages marked by rock formations, and a succession of new animals in shadowy conformity with these; but the rock formations and all the associated conditions make no stoppage or marked change at the time of man's appearance. He comes in the course of them, and goes—is still going along, in accordance with them. He is only a new guest, who has entered and sat down at a feast where other guests were before him, and which goes on and on continually: may there not be other guests to come and take their places at this perennial banquet of the High and Bountiful Master? Meaning by other guests, beings, not descending, (as common genealogical language would have it,) but *ascending*, from the now living Mankind,—possessing a superior development of the human character in accordance with the better external conditions which shall then have come into play,—favoured latter children of Nature, who have not lived till the throes and troubles of her maternal state were past. But is the improvement of these conditions to be left to the advance of physical nature, as that was seen before the existence of man? I suspect not. When man came upon the scene, a new agency was added to those formerly operating to this effect. By the work of our thoughtful brains and busy hands, we modify external nature in a way never known before. Under the operations of tillage, of mechanism, of building, making, and inventing; of those applications of natural powers and forces which human wit turns to account in so many ways; of all the results of social experience, of knowledge, and of arrangement; the earth tends to become a much serener field of existence than it was in the earlier ages of man's history. Its progress in this respect may not be clearly seen at a particular time, through the obscuring effect of temporary and accidental causes; but that the tendency of the physical improvements wrought by man upon the surface, and of the mechanic movements which he invents for the saving of his own labour, is to improve the daily comforts, and allow room for the intellectual and moral advancement of earth's children, cannot be denied without something like flying in the face of Providence itself. These improvements, then, thus partly wrought out by the exertions of the present race, I conceive as at once preparations for, and causes of, the *possible development* of higher types of humanity,—beings less strong in the impulsive parts of our nature, physical nature giving less matter for that nature to contend with and subdue to its needs,—more strong in the reasoning and the moral, because there will be less of the opposite to give these marring or check,—more fitted for the delights of social life, because society will then present less to dread and more to love.

The history and constitution of the world have now been explained according to the best lights which a humble individual has found within the reach of his perceptive and reasoning faculties. We have seen a system in which all is regularity and order, and all flows from and is obedient to a divine code of laws of unbending operation. We are to understand from what has been laid before us, that man, with his varied mental powers and impulses, is a natural problem, of which the elements can be taken cognizance of by science, and that all the secular destinies of our race, from generation to generation, are but evolutions from a primal arrangement in the counsels of Deity. To many, at first sight, it is apt to appear as a dreary view of the divine economy of our world, as if it placed God at an immeasurable distance from his creatures, and left them without refuge or remedy from the numberless ills that "flesh is heir to," and which no one can hope altogether to escape. But, in reality, God may be presumed to be revealed to us in every one of the phenomena of the system, in the suspension of globes in space, in the degradation of rocks and the upthrowing of mountains, in the development of plants and animals, in each movement of our minds, and in all that we enjoy and suffer, seeing that, the system requiring a sustainer as well as an originator, he must be continually present in every part of it, albeit he does not permit a single law to swerve in any case from its appointed course of operation. Thus, we may still feel that He is the immediate breather of our life and ruler of our spirits, that we may, by rightly directed thought, come into communion with him, and feel that, even when his penal ordinances are enforced upon us, his hand and arm are closely about us. Nor is this all. It may be that, while we are committed to take our chance in a natural system of undeviating operation, and are left, with apparent ruthlessness, to endure the consequences of every collision into which we knowingly or unknowingly come with each of its regulations, there is a system of Mercy and Grace behind the screen of nature, towards which we stand in a peculiar class of relations, which is capable of compensating for all casualties endured here, and whose very largeness is what makes these casualties a matter of indifference to God. For the existence of such a system, the actual constitution of nature is indeed a powerful argument. The reasoning may proceed thus:—the system of nature assures us that benevolence is a leading principle in the Divine Mind. But that system is at the same time deficient in a means of making this benevolence of invariable operation. To reconcile this to the character of the Deity, it is necessary to suppose that the present system is but a part of a whole, a stage in a Great Progress, and that the Redress is in reserve. Another argument here occurs—the economy of nature, beautifully arranged and vast in its extent as it is, does not satisfy even man's idea of what might be; he feels that, if this multiplicity of theatres for the exemplification of such phenomena as we see on earth were to go on for ever unchanged, it would not be worthy of the Being capable of creating it. An endless monotony of human generations, with their humble thinkings and doings, even though liable to a certain improvement, seems an object beneath that august Being. But the mundane economy might be very well as a portion of some greater phenomenon, the rest of which was yet to be evolved. Our system, therefore, though it may at first appear at issue with other doctrines in esteem amongst mankind, tends to come into harmony with them, and even to give them support. I would say, in conclusion, that, even where the two above arguments may fail of effect, there may yet be a faith derived from this view of nature sufficient to sustain us under all sense of the imperfect happiness, the calamities, the woes, and pains of this sphere of being. For let us but fully and truly consider what a system is here laid open to view, and we cannot well doubt that we are in the hands of One who is both able and willing to do us the most entire justice. Surely in such a faith we may well rest at ease, even though life should have been to us but a protracted malady, or though every hope we had built on the secular materials within our reach were felt to be melting from our grasp. Thinking of all the contingencies of this world as to be in time melted into or lost in some greater system, to which the present is only subsidiary, let us wait the end with patience, and be of good cheer.

NOTE CONCLUSORY.

THUS ends a book, composed in solitude, and almost without the cognizance of a single fellow-being, as solely as possible for the purpose of improving the knowledge of mankind, and through that medium, their happiness. For reasons best to be appreciated by myself, my name is retained in its original obscurity, and, in all probability, will never be generally known. I do not expect that any word of praise which the work may elicit shall ever be responded to by me; or that any word of censure shall ever be parried or deprecated. *Sine me*—I may say so far more truly than did the lorn exile of the Euxine—it goes forth to take its chance of instant oblivion, or of a long and active course of usefulness in the world. Neither contingency can be of any importance to me, beyond the regret or the satisfaction which may be imparted by my sense of a lost or a realized benefit to my fellow-creatures. The book, as far as I am aware, is the first attempt to connect the natural sciences into a history of creation, and thence to eliminate a view of nature as one grand system of causation. As a first effort towards a task so much above ordinary aspirations, it must necessarily be in some measure crude and unsatisfactory, even overlooking errors of detail justly attributable to my defective knowledge. Yet I have thought that the time was come for attempting to weave a great generalization out of the truths already established, or likely soon to be so—not that these were to be held as absolutely sufficient for the perfect completion of such an object, but that it is well, at certain times, to make advances into the field of speculation, in order that a direction may be given for the acquisition of new facts. If my hypothesis shall appear to be in the main a reflection of natural truth, I anticipate that attention will be drawn to dubious points; observations will be made, and discussions will take place; and in the long run, we shall find we have made a movement, and that towards a settlement of some of the greatest questions affecting humanity.

My sincere desire in the composition of the book was to give what upon mature reflection I conceive to be the true view of nature, with as little vexatious collision as possible with existing beliefs, whether philosophical or religious. I have made little reference to any doctrines of the latter kind which may be thought inconsistent with mine, because to do so would have been to enter upon questions for the settlement of which our knowledge is not yet ripe. Let the reconciliation of whatever is true in my views with whatever is true in other systems come about in the fulness of calm and careful inquiry. I cannot but here remind the reader of what Dr. Whewell has remarked in his History of the Inductive Sciences, and Dr. Wiseman illustrated strikingly in his lectures, how different new philosophic doctrines are apt to appear after we have become familiar with them. Geology at first seems inconsistent with the authority of the Mosaic record. A storm of unreasoning indignation rises against its teachers. In time, its truths, being found quite irresistible, are admitted, and mankind continue to regard the Scriptures with the same respect as before. So also with several other sciences. Now the only objection that can be made on such ground to this book, is, that it brings forward some new hypotheses, at first sight, like geology, not in perfect harmony with that record, and arranges these, with some associated facts, into a system which partakes of the same character. But may not the sacred text, on a liberal interpretation, or with the benefit of new light reflected from nature, or derived from learning, be shown to be as much in harmony with the novelties of this volume as it has been with geology and natural philosophy? What is there in the laws of organic creation more startling to the candid theologian than in the Copernican system or the natural formation of strata? And if the whole series of facts is true, why should we shrink from inferences legitimately flowing from it? Is it not a wiser course, since reconciliation has come in so many instances, still to hope for it, still to go on with our new truths, trusting that they also will in time be found harmonious with all others? Thus we avoid the damage which the very appearance of an opposition to natural truth is calculated to inflict on any system presumed to require such support. Thus we give, as is meet, a respectful reception to what is revealed through the medium of nature, at the same time that we fully reserve our reverence for all we have been accustomed to hold sacred, not one tittle of which it may ultimately be found necessary to alter.

APPENDIX.

*** It has been thought proper to remove to this place certain notes relating to controverted points, in order that the perusal of the text may not be disturbed with petty and, as I believe, temporary cavils. For more abundant answers to objections of every kind, general reference is made to *Explanations; a Sequel to Vestiges, &c.*

NOTE A, p. 6.

The six-feet reflector, erected by the Earl of Rosse, near Cork, and first brought into operation in the latter part of 1844, resolved a great number of the nebulæ which had resisted other instruments of inferior power. Upon this, it has been rashly assumed, in some quarters, that all the celestial objects, usually called nebulæ, were proved to be clusters of stars, rendered cloudlike only by the vast distance at which they were placed. The truth appears to be, that the nebulæ now resolved were of that class which were always considered as liable to be ascertained as *astral systems;* while another class of objects, described in pp. 7 and 8 of the text, remained unaffected by these observations. See *Explanations*, pp. 8—13.

Note B, p. 75.

While the cartilaginous fishes which now exist reach lower down in the scale of organization than the osseous, and while their imperfect vertebral structure, heterocercal tails, and other peculiarities, indicate a general inferiority, some of them present characters in the nervous and reproductive systems, which the osseous fishes do not possess. A few are viviparous, and manifest an affection for their offspring. On these partial grounds, an assumption has been built that the fishes commence with the highest forms. The occurrence of cestraceons in the Upper Silurians is particularly insisted upon as evidence for this conclusion. In reality, the few traits of superiority in the cartilaginous order, even if general to it, which they are not, are light in the scale, against the truly general inferiority. It is well known that no family of the animal kingdom is equally high in all points of structure and endowment, and that many forms, generally humble, have characteristics of a comparatively elevated kind. There are features of even the human organization which would place our race below some of the inferior animals, if these were to be made an exclusive criterion. The partial superiority possessed by several cartilaginous genera seems partly to relate to their place in creation as destructives: they have a well-developed nervous system to enable them to conquer their prey, (see *Explanations*, pp. 49—56.) That the nervous system determines the character of the reproductive system is an admitted law in physiology, (see *Owen*, *Philosophical Transactions*, 1834, p. 359.) To find, then, some of these cartalagines, exhibiting a generative system superior to other fishes, is no true difficulty in our course. On the very same ground, the star-fishes, (radiata,) where the sexes are in different individuals, are superior to the annelides, (articulata,) which present "an androgynous combination of simple ovaria and testes;" yet no one would think of describing the radiata generally as superior to the articulata Or the polypes might be said to be superior to the star-fishes, because in some of them "the digestive canal presents an œsophagus, a gizzard, a glandular stomach, and an intestine," while the latter animals have only "a radiated sac with one aperture." Yet, does any one, for that reason, think of placing the polypes above the star-fishes? It cannot be pretended that these and many similar facts are not well known, for they are in every tolerable manual of physiology. Yet, in direct contradiction of them, the opponents of the theory of development persist in asserting that the first fishes in the geological record are the highest in the book of the zoologist.

While insisting on the general inferiority of the cartilaginous to the osseous fishes, we must observe that the development of the animal kingdom appears to have been in distinct lines, the line of the carnivorous animals being abreast of the others, so that an early appearance of cestraceons, in comparison with other cartilaginous fishes, ought not to be surprising. For further explanations on this point, the reader is referred to the chapter of the present volume, entitled *The Affinities and Geographical Distribution of Organisms.*

NOTE C, p. 86.

From the experiment of Professor Lindley, which seemed to prove that dicotyledonous trees perish in water sooner than the monocotyledons, it has been said that, probably, we only find the carbonigenous vegetation to be lowly because the higher trees were incapable of being preserved. It is, however, remarkable, that the dicotyledons abound in the tertiary strata, which could hardly have been the case if they were incapable of resisting the effects of water. The objectors would need, at least, to account for these trees withstanding dissolution in that age, if they are to be supposed to have perished so readily in the earlier epoch. It is also to be remarked, that the dicotyledons do exist in the carbonigenous era; only they are extremely few. Finding simple sea-plants in the earlier fossiliferous strata and dicotyledons abundant in the last, while the intermediate carbonic period presents the intermediate kinds of plants in abundance, and only a scantling of any higher forms, it appears the most legitimate inference in the case, that the earth has witnessed a botanical progress connected with time, and only reached the highest vegetable forms at a comparatively recent period; thus presenting a history entirely analogous to what geology shows us of the animal kingdom.

Note D, p. 93.

The early occurrence of fishes, with a peculiarity of structure allying them to the reptilian class, while fishes possessing no reptilian affinities come into existence, in large numbers, long afterwards, is sometimes brought forward as one of the proofs that the fish class commenced with its highest forms. In strict fact, the Sauroids are not the first fish: they were preceded in the Upper Silurian formation by Placoids, and in the chart of M. Agassiz, (copied in Jameson's Journal, Oct. 1844,) they come after another large family of their own order, the Lepidoids. With regard to the subsequent rise of non-reptilian fishes, the reader will see some suggestions in the chapters on *The Affinities and Geographical Distribution of Organisms.*

This may be the best place at which to make reference to certain reptilian remains recently found in strata, supposed to be of the New Red Sandstone, in South Africa. One portion of these remains indicates an animal more huge than the crocodile. Another goes to form a new lacertian genus, combining characters of the lizard, crocodile, and tortoise, and to which Mr. Owen has given the name of *Dicynodon,* on account of two canine tusks which projected downwards with an outward curve from the upper jaw of the animal, the rest of the mouth being horny and toothless. These tusks, both as to their form and internal structure, are regarded as of mammalian character.

Here, too, it is said by the opponents of the development theory, we find traits of superior organization in the earliest animals of a particular class.

That these Bidentals, as Mr. Owen more comprehensively calls them, are amongst the earliest reptiles, is by no means ascertained; for the situation of the strata, in which they have been found, is unfixed. But, admitting that they were of early occurrence among reptiles, their exhibiting an approximation to mammalian dentition cannot truly be regarded as a proof of their being high in their class. We know well that a superior development of one organ, more especially an external one, tells nothing to that effect. The echinus, a member of the echinodermata, is furnished with teeth, while, in the superior family of holothuria, they are reduced to rudiments. Müller detected in the scorpion most of the parts which enter into the eye of the vertebrated animal, as well as a similarity in their arrangement, and yet we know how far inferior the scorpion is, on general grounds, to the vertebrate sub-kingdom. The fact is, that animals are endowed with such partial superiorities, when necessary with regard to the circumstances in which they are destined to live; but their place in the animal scale is to be determined on totally different considerations. How, if it were otherwise, should we find teeth in certain radiata, and wanting in the great bulk of the mollusca and articulata? How should we find this branch of organization, which prevails generally in the reptiles, become extinguished in the superior class of the birds, and even in some of the mammalia (the Manatus Stelleri, for example?)

In plain truth, the seizing upon this fact of bidental reptiles as a proof against the development theory, and that before even the place of the strata in which they were found was determined, is only an evidence of the rashness of the counter-theorists on this question, and of the weakness of the arguments by which their opposition is maintained.

Note E, p. 178.

"——— the form of the route of free electricity is modified by the medium through which it passes, and also by the electrical state of such medium, or of that of the relative electrical conditions of two bodies between which it is transmitted. If the medium through which it passes possesses a very inferior conducting power, it is obvious that a certain momentum must be requisite to enable the fluid to force its passage to a given distance, and there will be a point at which the momentum of the fluid and the resistance of the body will exactly counterbalance each other; but so soon as the electricity has again accumulated to a sufficient degree to overcome the resistance, it will again force its way in another direction, until it arrives at another point of equilibrium. In this way we may readily see the modus operandi of the electric fluid in imparting regular forms to bodies; and it is highly probable that its action in this respect extends to the vegetable kingdom, and perhaps operates even on animals, from the time in which they exist in the embryo state. . . . . Another fact, in support of the opinion that the distinctive forms of bodies are produced by electrical action, is, that crystals, and the twigs and leaves of vegetables, all terminate in points or sharp edges, so that the electrical action can proceed no further in increasing the growth, or, in other words, in propelling fresh portions of matter for the extension of the plant, or the crystal, beyond the pointed or edged termination."—*Leithead's Electricity*, 1837; p. 234.

Note F, p. 183.

The reader will please to understand that this is only a humble attempt to bring illustration from a department of science on which at present much doubt and obscurity rest. I have followed the best lights that could be found, but cannot be assured that better will not yet be evolved from the researches of the many able physiologists now engaged in the investigation of ultimate structure and of embryology. I am bound to admit, in the meantime, that the identity of the globules produced in albumen by electricity with *living cells*, and the fact of the reproduction of living globules, are both doubted by physiologists of high character. In this, as in other instances, particular illustrations may be held in doubt, or may altogether fail, without necessary injury to other arguments.

Note G, p. 189.

A more general, but more arresting argument in favour of primitive production, though not conclusively so, has been presented in the following terms:—

"We see a simple germ—the nucleus of a cell—develop itself into a feeling, moving, thinking man, by drawing into itself, and combining into new forms, the particles of what we are accustomed to call inorganic matter. These new forms are caused, by the very act of combination, to manifest properties of a new and peculiar kind; and their actions constitute the life of the being. Hence we must attribute to all those substances, which are thus drawn from the inorganic into the organic mode of existence, *a latent capacity* for the latter;—just as we say that the oxygen, hydrogen, carbon, and nitrogen, which make up the organic substance termed muscular fibre, and which, in *that* state or mode of combination, possess certain vital properties, possess also *a latent capacity* for combining in that mode of aggregation termed crystalline, and for exhibiting the solubility, translucency, and other qualities of a salt (all of which are totally opposed to its vital properties, and cannot co-exist with them), when united into the form of cyanate of ammonia. If we were only acquainted with those elements as they exist in organic compounds, their transposition into a crystalline salt would be almost as marvellous to us as the opposite change is now. If this *latent organizability or vitality* be admitted (as we conceive logical proof to have been given that it must) as a property of a large proportion of what we call inorganic matter, is there any such wonderful difficulty in imagining that it may be brought into play in some other manner than by the agency of a pre-existing germ? We think not. But let further investigation and more extended experience decide."—*British and Foreign Medical Review*, January, 1845.

NOTE H, p. 196.

The writer of the critique upon this work in the *British and Foreign Medical Review*, after saying that "none of the easy solutions which have been offered of the difficult problem presented by the appearance of this acarus, can be admitted," proceeds to make a few remarks much to the above purpose; and adds—"Not the least curious part of its (the acarus's) history is the series of metamorphoses which it undergoes before quitting the solution; these being *entirely different from the very slight changes which other acari undergo after their emersion from the egg.* Further, we believe it may be positively asserted, that, in whatever mode these acari are first generated, *it is not from eggs*; since, after they have escaped from the solution, they live in the neighbourhood, and readily breed; and their eggs, which we have ourselves seen, are quite large enough to have been readily visible in the solution, had they existed there."

The metamorphoses here adverted to will perhaps go some way to satisfy those who have objected that the acarus, belonging, as it does, to the articulata, is too high an animal to have been produced otherwise than from ova.

I would, nevertheless, remark that the Acarus Crossii is only brought forward as one illustration, and in order that an hypothesis which I think has strong probabilities on its side may have the benefit of any doubts that can be instituted with regard to the production of this creature. The decision of the question against the conclusion here leant to, would still leave much sound illustration, and not in the least affect the *general* argument.

NOTE I, p. 233.

In the first four editions, I had at this place a table, showing the succession of orders of animals in the ascending series of rocks, in harmonious relation with the succession of orders, in a scale of the animal kingdom constructed by the late Dr. Fletcher, and published by him in his *Rudiments of Physiology*. The same harmonious relation was shown between these successions and the series of characters presented by the human brain in its embryotic state. Finding, however, that, notwithstanding explanations of the most guarded kind, misapprehensions unfavourable to the hypothesis of development were perpetually arising from this table, I have omitted it from the present edition.

NOTE J, p. 262.

In the fourth edition, I had expressed a doubt of the cephalopoda passing into the cyclostomous fishes, as these seemed to be connected with another invertebrate order; reconsideration now induces me to adopt this connexion, as, for several reasons, the most likely.

ERRATUM.

In page 58, for *orthis*, read *orthoceras*.

T. C. Savill, Printer, 4, Chandos-street, Covent-garden.

EXPLANATIONS.

EXPLANATIONS:

A Sequel to

"VESTIGES OF THE NATURAL HISTORY OF CREATION."

BY THE AUTHOR OF THAT WORK.

SECOND EDITION.

LONDON:
JOHN CHURCHILL, PRINCES STREET, SOHO.
MDCCCXLVI.

CONTENTS.

PAGE

Design of the *Vestiges* explained 3

Proper Position of the Nebular Hypothesis in the Argument . 5

Imputed Failure of the Hypothesis from the Earl of Rosse's discoveries, denied 8

Experiments illustrating and confirming the Hypothesis by Professor Plateau 14

Objection from the retrogression of Uranus's Satellites considered 18

Objection respecting the convergence of atoms to a central nucleus, answered 19

The Nebular Hypothesis not a supersession of Deity, but only a description of his mode of working 24

Quetelet's inquiries, establishing law in mental operations . . 25

Limits of the system being under law, the whole is probably so 26

Question of the Origin of Organic nature 27

Geology proves it to have observed a progress in time . . . 31

Objections respecting this progress 32

Lower Silurian Fossils 33

Upper Silurian Fossils 47

Old Red Sandstone 49

PAGE

Carboniferous System 59
Permian System 63
Outline of the Genetic Plan of the Animal Kingdom 69
Bearing of this Plan on the Arguments of Objectors 76
Reptiles of the Muschelkalk, Lias, &c. 82
Objections as to first Footmarks of Birds 85
Objections as to Earliest Mammalia 87
Tertiary Formation 90
Opinions of Cuvier and Agassiz 99
Apology of Mr. Sedgwick for Over-Ardent Generalizations . . 101
Physiological Objections of Dr. Clark, of Cambridge 103
Views of others respecting Embryotic Development 106
Germs not alleged to be identical 109
Transmutation of Plants 111
Species a Term, not a Fact 113
Instances of Transmutation 115
Transmutation does not imply extinction of Elder Species . . 117
The Broomfield Experiment 119
Proof of Aboriginal Life in the present era not essential to the theory of Organic Creation by Law 121
The Opposite Theory characterized 125
Views of Dr. Whewell, and objections to them 127
Views of the Edinburgh Reviewer—these analyzed 135
Views of Professor Agassiz 139
Views of Sir John Herschel 141
Support to Theory of Law from Rev. Dr. Pye Smith and Blackwood's Magazine 143
Mr. Stuart Mill on Universal Causation 145
Present State of Opinion on the Origin of Organic Nature examined 149
Animals have not come immediately on the occurrence of proper conditions 151

PAGE

Great number of distinct Floras 152

Supposed Formation of New Species, as upheld by Professor Owen, &c., inadmissible 154

Opinions of Professor Pictet on Peculiarity of Species in each formation 155

Time the true key to difficulties arising from apparent permanency of species 158

Vast spaces of time involved in the geological record 159

Zoology of Galapagos Islands, an instance of comparatively recent development 161

Author's theory supported by facts connected with the distribution of plants 165

Whence the first impulse to vitality? 168

The *Vestiges*—its object purely scientific—defended on this ground 169

Ungenerous policy of Geological Objectors 170

Opposition of the Scientific Class 174

Estimate of this Opposition 175

Utility of Hypotheses 179

Bearing of the new doctrine on Human Interests 182

Its Moral Results 184

Consolations and Encouragements offered by it 187

APPENDIX—Letters of Mr. Weekes on Aboriginal Production of Insects 189

EXPLANATIONS.

WHEN the work to which this may be regarded as a supplement was published, my design was not only to be personally removed from all praise or censure which it might evoke, but to write no more upon the subject. I said to myself, Let this book go forth to be received as truth, or to provoke others to a controversy which may result in establishing or overthrowing it; but be my task now ended. I did not then reflect that, even though written by one better informed or more skilled in argument than I can pretend to be, it might leave the subject in such a condition that the author should have to regret seeing it in a great measure misapprehended in its general scope, and also so much excepted to, justly and unjustly, on particular points, that ordinary readers might be ready to suppose its whole indications disproved. Had I bethought me of such possible results, I might have announced, from the beginning, my readiness to enter upon such explanations of points objected to, and such reinforcements of the general argument, as might promise to be serviceable. And this would have seemed the more necessary, in as far as it may be expected that there are many points in a new and startling hypothesis which no one can be so well qualified to clear up and strengthen as its author. I might have felt, at the same time, that a new adventure, for whatever purpose, in the same field, was hazardous, with regard to any favourable impression previously produced; yet such an objection would, again, have been at once overruled, seeing that public favour and disfavour were alike beyond the regard of an author who bore no bodily shape in the eyes of his fellow-countrymen, and was likely to remain for ever unknown. Such reflections now occur to me, and I am consequently induced to take up the pen for the purpose of endeavouring to make good what is deficient, and reasserting and confirming whatever has been unjustly challenged in my book. In doing so, I shall study to direct attention solely to fact and argument, or what appear as such, overlooking the uncivil expressions which the work has drawn forth in various quarters, and which, of course, can only be a discredit to their authors.

I must start with a more explicit statement of the general argument of the *Vestiges*, for this has been extensively misunderstood. The book is not primarily designed, as many have intimated in their criticisms, and as the title might be thought partly to imply, to establish a new theory respecting the origin of animated nature; nor are the chief arguments directed to that point. The object is one to which the idea of an organic creation in the manner of natural law is only subordinate and ministrative, as likewise are the nebular hypothesis and the doctrine of a fixed natural order in mind and morals. This purpose is to show that the whole revelation of the works of God presented to our senses and reason is a system based in what we are compelled, for want of a better term, to call LAW; by which, however, is not meant a system independent or exclusive of Deity, but one which only proposes a *certain mode of his working.* The nature and bearing of this doctrine will be afterwards adverted to; let me, meanwhile, observe, that it has long been pointed to by science, though hardly anywhere broadly and fully contemplated. And this was scarcely to be wondered at, since, while the whole physical arrangements of the universe were placed under law by the discoveries of Kepler and Newton, there was still such a mysterious conception of the origin of organic nature, and of the character of our own fitful being, that men were almost forced to make at least large exceptions from any proposed plan of universal order. What makes the case now somewhat different is, that of late years we have attained much additional knowledge of nature, pointing in the same direction as the physical arrangements of the world. The time seems to have come when it is proper to enter into a re-examination of the whole subject, in order to ascertain whether, in what we actually know, there is most evidence in favour of an entire or a partial system of fixed order. When led to make this inquiry for myself, I soon became convinced that the idea of any exception to the plan of law stood upon a narrow, and constantly narrowing foundation, depending, indeed, on a few difficulties or obscurities, rather than objections, which were certain soon to be swept away by the advancing tide of knowledge. It appeared, at the same time, that there was a want in the state of philosophy amongst us, of an impulse in the direction of the consideration of this theory, so as to bring its difficulties the sooner to a bearing in the one way or the other; and hence it was that I presumed to enter the field.

My starting point was a statement of the arrangements of the bodies of space, with a hypothesis respecting the mode in which those arrangements had been effected. It is a mistake to suppose this (nebular) hypothesis essential, as the basis of the entire system of nature developed in my book. That basis lies in the material laws found to prevail throughout the universe, which explain why the masses of space are globular; why planets revolve round suns in elliptical orbits; how their rates of speed are high in proportion to their nearness to the centre of attraction; and so forth. In these laws arise the first powerful presumption that the formation and arrangements of the celestial bodies were brought about by the Divine will, *acting in the manner of a fixed order or law,* instead of any mode which we conceive of as more arbitrary. It is a presumption which an enlightened mind is altogether unable to resist, when it sees that precisely similar effects are every day produced by law on a small scale, as when a drop of water spherifies, when the revolving hoop bulges out in the plane of its equator, and the sling, swung round in the hand, increases in speed as the string is shortened. The philosopher, on observing these phenomena, and finding incontestable proof that they are precisely of the same nature as those attending the formation and arrangement of worlds, learns his first great lesson — that the natural laws work on the minutest and the grandest scale indifferently; that, in fact, there is no such thing as great and small in nature, but world spaces are as a hair-breadth, and a thousand years as one day. Having thus all but demonstration that the spheres were formed and arranged by natural law, the nebular hypothesis becomes important, as shadowing forth the process by which matter was so transformed from a previous condition, but it is nothing more; and, though it were utterly disproved, the evidence which we previously possessed that physical creation, so to speak, was effected by means of, or in the manner of law, would remain exactly as it was. We should only be left in the dark with regard to the previous condition of matter, and the steps of the process by which it acquired its present forms.

It would nevertheless strengthen the presumption, and, indeed, place it near to ascertained truths, if we were to obtain strong evidence for what has hitherto been called the nebular hypothesis. The evidence for it is sketched in the *Vestiges:* it is exhibited with greater clearness, and in elegant and impressive language, in Professor Nichol's *Views of the Architecture of the Heavens.* The position held by this hypothesis in the philosophical world when my book was written, is shown, with tolerable distinctness, in the *Edinburgh Review* for 1838, where it is spoken of in the following general terms:— "These views of the origin and destiny of the various systems of worlds which fill the immensity of space, break upon the mind with all the interest of novelty, and *all the brightness of truth.* Appealing to our imagination by their grandeur, and to our reason by the *severe principles on which they rest*, the mind feels as if a revelation had been vouchsafed to it of the past and future history of the universe." It may also be remarked that this writer considered the hypothesis as "confirming, rather than opposing the Mosaic cosmogony, whether allegorically or literally interpreted." With this testimony to the mathematical expositions of MM. La Place and Comte, I rest content, as the expositions themselves would be unsuitable in a popular treatise. But the hypothesis has been favourably entertained in many authoritative quarters, during the last few years, and probably would have continued to be so, if no attempt had been made to enforce by it a system of nature on the principle of universal order.

The chief objection taken to the theory is, that the existence of nebulous matter in the heavens is disproved by the discoveries made by the Earl of Rosse's telescope. By this wondrous tube, we are told, it is shown to be "an unwarrantable assumption that there are in the heavenly spaces any masses of matter different from solid bodies composing planetary systems."* The nebulæ, in short, are said to be now shown as clusters of stars, rendered apparently nebulous only by the vast distance at which they are placed. There is often seen a greater vehemence and rashness in objecting to, than in presenting hypotheses; and we appear to have here an instance of such hasty counter-generalization. The fact is, that the nebulæ were always understood to be of two kinds: 1, nebulæ which were only distant clusters, and which yielded, one after another, to the resolving powers of telescopes, as these powers were increased; 2, nebulæ comparatively near, which no increase of telescopic power affected. Two classes of objects wholly different were, from their partial resemblance, recognised by one name, and hence the confusion which has arisen upon the subject. The resolution of a great quantity of the first kind of nebulæ by Lord Rosse's telescope was of course expected, and it is a fact, though in itself interesting, of no consequence to the nebular hypothesis. It will only be in the event of the second class being also resolved, and its being thus shown

* North British Review, iii. 477.

that there is only one class of nebulæ, that the hypothesis will suffer. Such, at least, I conclude to be the sense of a passage which I take leave to transfer, in an abridged form, from a recent edition of Professor Nichol's work.

"I. By far the greater number of the milky streaks, or spots, whose places have hitherto been recorded, lie at the outermost, or nearly at the outermost boundary of the sphere previously reached by our telescopes: and in this case there is no certain principle on the ground upon which a pure nebula can be distinguished from a cluster so remote that only the general or fused light of its myriads of constituent orbs can be seen. Sometimes,—resting on a peculiarity of form or other characteristic,—the astronomer may venture a guess that such an object is probably a firmament; as, indeed, I was bold enough to do in former editions of this work with regard to several which have since been resolved; but, in the main, he can tell little concerning them, or have any other belief, than that, as with similar masses near him, a great, probably the greater number, are true clusters, grand arrangements of stars, incredibly remote, but resembling in all things our own home galaxy. Now, the application to such objects of a new and enlarged power of vision, could be attended only by one result—magnificent, but far from unexpected: and it is here that the six-feet mirror has achieved its earliest triumphs. Under its piercing glance, great numbers of the milky specks have unfolded their starry constituents; some of these, which previously were almost unresolved, shining with a lustre equivalent to that of our brightest orbs to the naked eye. How far it will go with its resolving power has not yet been ascertained; but I perceive that Sir James South has given his authority that some spots examined by it continue intractable.

"II. The influence of the new discoveries either to impair or strengthen the foundations of the nebular hypothesis, must clearly be looked for among their bearings on less remote and ambiguous objects. Now, the new aspects of these may lead us to question our former opinions as to the existence of the supposed filmy self-luminous masses,—or they may throw doubt on the reality of those forms according to which we have arranged them, and which seem to indicate the steps of a stupendous progress.

"1. Astronomers have never rested their belief in the reality and wide diffusion of the nebulous matter, on the objects referred to in the first paragraph; but on others, much within the range of our previous vision. In so far as we have hitherto understood the nature of clusters, the telescopic power required to resolve them is never very much higher than that which first descries them as dim milky spots. But there are many most remarkable objects which, in this essential feature, are wholly contrasted with clusters. For instance, the nebula in Orion, as I have fully shown in the text, is visible to the naked eye, as also is the gorgeous one in Andromeda; while the largest instrument heretofore turned to them has given no intimation that their light is stellar, but rather the contrary; although small stars are found buried amidst their mass. Now, if Lord Rosse's telescope resolves these, and others with similar attributes, such as some of the streaks among the following plates, we shall thereby be informed that we have generalised too hastily from the character of known firmaments,—that schemes of stellar being exist, infinitely more strange and varied than we had ventured to suppose,—and certainly we shall then hesitate in averring further, concerning the existence or at least the diffusion of the purely nebulous modification of matter.

"2. Lord Rosse's telescope may also, as I have said, disprove the reality of our arrangement of the forms of the nebulæ as steps of a progression. And in regard of this question, there seem two classes of objects meriting attention.

"*First*, I shall refer to the nebulous stars properly so called, or to that form in which the diffused matter has reached the condition of almost pure fixed stars. Now, of these objects there are two distinct sets, presenting at first to the telescope very much the same appearance, but in regard of which our knowledge is very different. It will readily be conceived that a distant cluster, with strong concentration about the centre of its figure, must, to the telescope which first descries it, look like a star with a halo around it. When a higher power is applied, that central star, however, will appear as a disc, and to a still higher power the cluster will be revealed. A very great number of what are called nebulous stars, are doubtless of this class; and we have hitherto had no means of accurately ascertaining the fact, just because our largest telescopes were required to descry them; but there are multitudes of others—the true 'photospheres'—quite of a different description. Many of these are easily seen as fixed stars with haloes of different sizes diminishing to the mere 'bur;' and under the greatest power as yet applied, the apparent central star never expands into a disc, or departs from the stellar character. It is by its effect on these that the new instrument will at all bear on this portion of the nebular hypothesis.

"*Secondly*, The foregoing being our grounds of belief in the existence of nebulæ—first, in a diffused or chaotic state, and again in a condition proximate to pure stars; the only remaining point has reference to nebulæ in an intermediate state,—when the roundish masses seem to have begun a process of organization or concentration, and carried it onwards through several stages:—a state to which we have every variety of analogon in the various forms and densities of cometic nuclei. Sir William Herschel certainly was not ignorant that round or spherical clusters abound in the skies, which, when first seen, present all the appearances of such nebulæ—nay, he grounded on the fact of their approximate sphericity and varying degrees of concentration, some of the boldest and most engrossing of his conjectures; nor would he have doubted that multitudes which, even to his instruments, seemed only general lights, would, in after times, be resolved; but here, as before, the gist of the question is not, can you resolve round nebulæ never resolved before; but can you resolve such as, quite within the range of former vision, have continued intractable under the scrutiny of powers which, judging from the average of our experience, must surpass what ought to have resolved them?

"Such are my views as to the present condition of this important question; and if they are correct, it will appear that, notwithstanding the resolutions achieved by the new instruments, they are, as yet, quite as likely—by accumulating new objects belonging to the three foregoing classes, and by more surely and distinctly establishing their characteristic features—to strengthen, as to invalidate the grounds of the nebular hypothesis. Eagerly, but patiently, let us watch the approaching revelations."

Various minor objections have been presented to the nebular hypothesis; but, before adverting to any of them, I may give a brief abstract of certain recent experiments, by which it has been remarkably illustrated. Here it is peculiarly important to bear in mind, that the phenomena of nature are, if I may so speak, indifferent to the scale on which they act. The dew-drop is, in physics, the picture of a world. Remembering this, we are prepared, in some measure, to hear of a Belgian professor imitating the supposed formation and arrangement of a solar system, in some of its most essential particulars, on the table of a lecture-room! The experiments were first conducted by Professor Plateau of Ghent, and afterwards repeated by our own Dr. Faraday.

The following abstract of Professor Plateau's experiments is also presented in the fifth edition of the *Vestiges.* Its being repeated here is, that it may meet the eyes of many who are not likely to see any edition of that work besides those from which it is absent:

Placing a mixture of water and alcohol in a glass box, and therein a small quantity of olive oil, of density precisely equal to the mixture, we have in the latter *a liquid mass relieved from the operation of gravity*, and free to take the exterior form given by the forces which may act upon it. In point of fact, the oil instantly takes a globular form by virtue of molecular attraction. A vertical axis being introduced through the box, with a small disc upon it, so arranged that its centre is coincident with the centre of the globe of oil, we turn the axis at a slow rate, and thus set the oil sphere into rotation. "We then presently see the sphere *flatten at its poles* and *swell out at its equator*, and we thus realize, on a small scale, an effect which is admitted to have taken place in the planets." The spherifying forces are of different natures, that of molecular attraction in the case of the oil, and of universal attraction in that of the planet, but the results are "analogous, if not identical." Quickening the rotation makes the figure more oblately spheroidal. When it comes to be so quick as two or three turns in a second, "the liquid sphere first takes rapidly its maximum of flattening, then becomes hollow above and below, around the axis of rotation, stretching out continually in a horizontal direction, and finally, abandoning the disc, is *transformed into a perfectly regular ring*." At first this remains connected with the disc by a thin pellicle of oil; but on the disc being stopped this breaks and disappears, and the ring becomes completely disengaged. The only observable difference between the latter and the ring of Saturn is, that it is rounded instead of being flattened; but this is accounted for in a satisfactory way.

A little after the stoppage of the rotatory motion of the disc, the ring of oil, losing its own motion, gathers once more into a sphere. If, however, a smaller disc be used, and its rotation continued after the separation of the ring, rotatory motion and centrifugal force will be generated in the alcoholic fluid, and the oil ring, thus prevented from returning into the globular form, divides itself into "*several isolated masses, each of which immediately takes the globular form.*" These are "almost always seen to assume, at the instant of their formation, a *movement of rotation upon themselves* — a movement which constantly takes place *in the same direction as that of the ring.* Moreover, as the ring, at the instant of its rupture, had still a remainder of velocity, the spheres to which it has given birth tend to fly off at a tangent; but as, on the other side, the disc, turning in the alcoholic liquor, has impressed on this a movement of rotation, the spheres are especially carried along by this last movement, and revolve for some time round the disc. Those which revolve at the same time upon themselves, consequently, then present the curious spectacle of *planets revolving at the same time on themselves and in their orbits.* Finally, another very curious effect is also manifested in these circumstances: besides three or four large spheres into which the ring resolves itself, there are almost always produced one or two very small ones, which may thus be compared to satellites. The experiment which we have thus described presents, as we see, an image in miniature of the formation of the planets, according to the hypothesis of Laplace, by the rupture of the cosmical rings attributable to the condensation of the solar atmosphere."*

Such illustrations certainly tend to take from the nebular cosmogony the character of a "splendid vision," which one of my critics has applied to it. I may here also remind the reader that there are other grounds for this hypothesis, besides observations on the nebulæ. Overlooking the

* Pr. Plateau on the Phenomena presented by a free Liquid Mass withdrawn from the action of gravity. Taylor's Scientific Memoirs. November, 1844.

zodiacal light, which has been thought a residuum of the nebulous fluid of our system, we find geology taking us back *towards* a state of our globe which cannot otherwise be explained. It was clearly at one time in a state of igneous fluidity,—the state in which its oblately spheroidal form was assumed under the law of centrifugal force. Since then it has cooled, at least in the exterior crust. We thus have it passing through a chemical process attended by diminishing heat. Whence the heat at first, if not from the causes indicated in the nebular hypothesis? But this is not all. In looking back along the steps of such a process, we have no limit imposed. There is nothing to call for our stopping till we reach one of those extreme temperatures which would vaporize the solid materials; and this gives us exactly that condition of things which is implied by the nebular cosmogony.

Of particular objections it is not necessary to say much. That there should be difficulties attending such an hypothesis is only to be expected; but where general evidence is so strong, we should certainly be scrupulous about allowing them too much weight. It is represented, for instance, that the matter of the solar system could not, in any conceivable gaseous form, fill the space comprehended by the orbit of Uranus. If this be the case, let it be allowed as a difficulty. It is pointed out that the planets do not increase regularly in density from the outermost to the innermost. Their sizes are also not in a regular progression, though the largest, generally speaking, are towards the exterior of the system. It was not, perhaps, to be expected, that such gradations should be observed; but, grant there was some reason to look for them, their absence constitutes only another and a slight difficulty. Then we know no law to determine the particular "stages at which rings are formed and detached." Be it so—although something of the kind there doubtless is, as the distances of the planets, according to Bode's law, observe a geometrical series of which the ratio of increase is 2. From these objections, which cannot now be answered, let us pass to some which can.

It has been said that a confluence of atoms towards a central point, as presumed by the nebular hypothesis, would result, not in a rotation, but in a state of rest.* According to the *North British Review*—". . .Supposing the uniformly distributed atoms to agglomerate round their ringleader, the space left *blank* by the slow advance of the atoms in radial lines converging to the nucleus, must be a ring bounded by concentric circles, the outermost circle being the limit of the nebulous matter not drawn to the centre of the nascent sun. Now, as all the forces which act upon the agglomerating particles, whether they proceed from the circumference of the undisturbed nebulous matter, or from the gradually increasing nucleus, must have their resultants in the radial lines above mentioned,—there can be no cause whatever capable of giving a rotatory motion to the mass. It must remain at rest."

Now, there can be no doubt that a confluence proceeding precisely to a centre, has this result; but this is only an abstract truth, not an exact and absolute description of any actual confluence of the kind. The explanation was afforded by Professor Nichol, long before the objection was started, and it could not be given in better

* North British Review, No. 6. Atlas Newspaper, Aug. 30, 1845.

language on the present occasion: "When we reflect on the solar nebula in the act of condensing, it appears that the act consists in a flow or rush of the nebulous matter from all sides towards a central region; which is virtually equivalent, in a mechanical point of view, to what we witness so frequently, both on a small and large scale—the meeting and intermingling of opposite gentle currents of water. Now, what do we find on occasion of such a meeting? Herschel's keen glance lighted at once on this simple phenomenon, and drew from it the secret of one of the most fertile processes of Nature! *In almost no case do streams meet and intermingle, without occasioning, where they intermingle, a dimple or whirlpool; and, in fact, it is barely possible that such a flow of matter from opposite sides could be so nicely balanced in any case, that the opposite momenta or floods would neutralize each other, and produce a condition of central rest.* In this circumstance, then—in the whirlpool to be expected where the nebulous floods meet—is the obscure and simple germ of rotatory movement. The very act of the condensation of the gaseous matter as it flows towards a central district, almost necessitates the commencement of a process, which, though slow and vague at first, has, it will be found, the inherent power of reaching a perfect and definite condition . . ."*

The exception presented by the satellites of Uranus to the otherwise uniform orbitual movements of the planetary bodies, is brought forward as a startling difficulty.† It is, in reality, only a trifling objection, seeing that so many other movements follow one rule, and that we may any day be able to fix upon a cause for this exception, perfectly in harmony with all the associated facts. There was once a similar difficulty in geology—strata uppermost where they ought to have been lowermost; but it was in time cleared. Geologists found that there had been a folding over of the strata, so as to reverse their proper and original positions. May we not rest in hope, that a similar exception in astronomy may find a similar solution? I have thrown out the hint of a possible *bouleversement* of the whole of that planet's system: it has been scoffed at; but it is only the sup-

* Views of the Architecture of the Heavens. First edition, 1837.

† Edinburgh Review, No. 165, p. 24.

position of a greater degree of obliquity in the inclination of the axis of the planet to the plane of its orbit than what we find in several others. The same causes which made the inclination of the axis of Venus towards her orbit 75 degrees, may have turned that of Uranus a little further along, and so reversed the position of his poles. The admitted inclination of the axis of Uranus towards the plane of his orbit is 79 degrees, being the greatest found in any of the planets. This implies only the necessity for an increase of inclination to the extent of 22 degrees, or about one-fourth of the quadrant, in order to account for the surmised reverse arrangement. Nor are causes for such a phenomenon far to seek. In the revolution of the presumed nebular mass, there would be great undulations, as I venture to say there would be found in any similar body which we might set into a similar rotatory motion. Such I esteem as the causes of the departure of the planetary axes from the vertical. A curve in the outermost portion, amounting to a fold—like the curl of a high wave — would cause the *bouleversement* of Uranus, and the consequent (apparent) retrogression of his satellites.

It appears, then, that, overlooking a few minor unexplained difficulties, the objections to the nebular hypothesis are not formidable to it. It approaches the region of ascertained truths, and may reasonably be held as a strong corroboration of what first appears from the material laws of the universe, that the whole Uranographical arrangements were effected in the manner of natural law. It is, however, altogether a mistake to regard this conclusion, as far as it is one, as equivalent to a superseding of Deity in the history of creation. It proposes nothing beyond a view of the mode in which the Divine Will has been pleased to act, in this first and most important of its works. The formation of worlds and their arrangement now appear but as steps in an Historical Progress, for matter is necessarily presumed to have existed before in a different form. By what means and under what circumstances creation, in the true sense of the word, took place, —that is, how existence was given to the matter which we suppose to have been capable of such evolutions—no one can as yet tell; we only are sure, if any trust can be placed in the laws of our minds, that it had a Cause, or an Author. Leaving such an inquiry as one, in which we have not, at present, ground for a single step, it is surely a great gratification that we can at least trace the operations of the Great First Cause, from a condition of matter anterior to its present forms, and learn with certainty that these operations were in no way arbitrary or capricious, that they were not single and detached phenomena, but the result of principles flowing from the Eternal and Immutable, and which prevailed over all the realms of Infinity at once.

We have fixed mechanical laws at one end of the system of nature. If we turn to the mind and morals of man, we find that we have equally fixed laws at the other. The human being, a mystery considered as an individual, becomes a simple natural phenomenon when taken in the mass, for a regularity is observed in every peculiarity of our constitution and every form of thought and deed of which we are capable, when we only extend our view over a sufficiently wide range. It is to M. Quetelet, of Brussels, that we are indebted for the first satisfactory explication of this great truth: it is presented in his well-known and very able treatise *Sur L'Homme, et le Développement de ses Facultés.* He first shows the regularity which presides over the births and deaths of a community, liable to be affected in some degree by accidental circumstances, but fixed again when these are uniform. He then makes it clear that the stature, weight, strength, and other physical peculiarities of men are likewise regulated by fixed principles in nature. Afterwards, the moral qualities,—the impulses of all our various sentiments and passions,—even the tendency to yield to those temptations which give birth to crime,—are proved to be of no less determinate character, however impossible it may be to predict the conduct of any single person. These are doctrines not to be resisted by inconsiderate prejudices. They rest on the most powerful of all evidence, that of numbers. If they appear to take from the personal responsibility of individuals, it is merely an appearance, for the doctrine immediately steps forward to show that laws, education, and moral influences of every kind exercise an equally determinate control over men; so that the need for their being called into use becomes even more palpable than before. We are not, however, required at this moment to argue respecting the bearing which this doctrine may have upon human interests. What we are at present concerned with is the simple fact, that Morals—that part of the system of things which seemed least under natural regulation or law—is as thoroughly ascertained to be wholly so, as the arrangements of the heavenly bodies.

Now we have here two most remarkable truths. The wondrous masses which people the Mighty Void are under the control of natural law. The workings of the little world of the human mind—the opposite extreme of the system—are under law likewise. We have thus the character of the *limits* of the system fixed. So far we proceed upon solid ground. Now it has been seen that phenomena precisely the same as the formation and arrangement of worlds take place daily before our eyes, under the influence of the laws of matter, showing that the whole cosmogony might have been effected—proving, indeed, that it *was* effected—by the Divine will acting in that manner. Having attained this point, we are called upon to remember the many appearances of unity in nature; how, when we take a sufficiently wide view, there is nothing discrepant and exceptive in it; how a noble and affecting simplicity breathes from it in every part. So reflecting, we ask, "Can it be that, as the first and the last parts of the system are under law, and the first (this being also the greatest) was manifestly created in that manner, so the whole is under law, and has been produced in that manner?" It is at the moment when we have arrived at this question, that the origin of the organic world becomes a point of importance. The sceptic of science steps in, and says, "No; the idea of an entire system under law, and produced by it, here breaks down, for who can pretend to penetrate the mysteries of vitality and organization? and who can say that species have had other than a miraculous origin?" The tone in which this objection is usually made seems to me inappropriate, considering that the objectors stand on a mere fragment of nature, and one which the discoveries of science are every day lessening. It is but in a nook, to which light has not yet fully penetrated, that the opponents of the theory of universal order take refuge. On coming to the consideration of the question, I am at the very first struck by the great *à priori* unlikelihood that there can have been two modes of Divine working in the history of nature—namely, a system of fixed order or law in the formation of globes, and a system in any degree different in the peopling of these globes with plants and animals. Laws govern both: we are left no room to doubt that laws were the immediate means of making the first; is it to be readily admitted that laws did not preside at the creation of the second also, particularly when we find that laws equally at this moment govern and sustain both? Most undoubtedly, it would require very powerful evidence to justify such an admission. And, on the other hand, it would require very decisive counter-evidence to forbid the conclusion that the organic creation originated in law. How actually stands the evidence on either side? Simply thus: that no actual evidence has ever yet been offered to prove that the Divine will acted otherwise than in the usual natural order in the organic creation; while, on the other hand, geology and physiology exhibit *lively vestiges or traces of that mode having actually been followed.* On this narrow ground, it appears, is the great question to be debated. If the opponents of the hypothesis of an organic creation by law can bring, from these or any other sciences, facts which appear as powerful objections to any such conclusion, then it must, at the very least, be held in suspense. If, again, the other party can show these sciences as presenting far more argument for a law-creation of organisms than against it, the hypothesis must be admitted to have the advantage. I have so presented these sciences; the evidence has been disputed, and some obscure points have been largely insisted upon in objection. It is now my duty to enter into the consideration of these objections, and see if they are really of the importance which has been attributed to them.

Fifty years ago, science possessed no facts regarding the origin of organic creatures upon earth; as far as knowledge acquired through the ordinary means was concerned, all was a blank antecedent to the first chapters of what we call ancient history. Within that time, by researches in the crust of the earth, we have obtained a bold outline of the history of the globe, during what appears to have been a vast chronology intervening between its formation and the appearance of the human race upon its surface. It is shown, on powerful evidence, that, during this time, strata of various thickness were deposited in seas, each in succession being composed of matters worn away from the previous rocks; volcanic agency broke up these strata, and projected chains of mountains; sea and land repeatedly changed conditions; in short, the whole of the arrangements which we see prevailing in the earth's crust took place, and that most undoubtedly under the influence of natural laws which we yet see continually operating. The remains and traces of plants and animals found in the succession of strata, show that, while these operations were going on, the earth gradually became the theatre of organic being, simple forms appearing first, and more complicated afterwards. *A time when there was no life* is first seen. We then *see life begin, and go on;* but whole ages elapsed before man came to crown the work of nature. This is a wonderful revelation to have come upon the men of our time, and one which the philosophers of the days of Newton could never have expected to be vouchsafed. The great fact established by it is, that the organic creation, as we now see it, was not placed upon the earth at once; —it observed a PROGRESS. Now we can *imagine* the Deity calling a young plant or animal into existence instantaneously; but we see that he does not usually do so. The young plant and also the young animal go through a series of conditions, advancing them from a mere germ to the fully developed repetition of the respective parental forms. So, also, we can *imagine* Divine power evoking a whole creation into being by one word; but we find that such had not been his mode of working in that instance, for geology fully proves that organic creation passed through a series of stages before the highest vegetable and animal forms appeared. Here we have the first hint of organic creation having arisen in the manner of natural order. The analogy does not prove identity of causes, but it surely points very broadly to natural order or law having been the mode of procedure in both instances.

But the question is, Does geology really show such a progress of being? This has been denied in some quarters, and particularly in the elaborate criticism upon the *Vestiges*, which appeared in the *Edinburgh Review*.* In reality the whole of the geologists admit that we have first the remains of *invertebrated animals;* then with these, *fish*, being the lowest of the vertebrated; next, *reptiles and birds*, which occupy higher grades; and, finally, along with the rest, *mammifers*, the highest of all; and yet controversialists will be found gravely telling their readers, "It is not

* July, 1845.

true that only the lowest forms of animal life are found in the lowest fossil bands, and that the more complicated structures are gradually developed among the higher bands, in what we might call a natural ascending scale;"* the pretext for giving this unqualified contradiction to the above grand fact being, that when we take the *special groups of animals*, as the invertebrata, the fishes, the reptiles, &c., there are some real or apparent grounds for denying that the low forms *of these groups* came before the higher. The fallacy consists in sinking the great broad palpable facts of the case, about which not the least doubt anywhere exists, and giving prominence to certain facts of far inferior magnitude, and comparatively obscure, but in whose obscurity there is a possibility of creating a kind of diversion. I trust to be able to show that, even in the special groups of fossils, there is no real obstacle to the theory of a gradual natural development of life upon our planet.

The view which the Edinburgh critic gives of the earliest stratified rocks is much the same as my own account of them. There is a *Hypozoic formation*, or series, devoid of remains of plants and animals; then a formation (*Lower Silurian*),

* "Edinburgh Review."

called in my early editions, The Clay-slate and Grawacke system, in which we find "no animals of the higher classes, with a regular skeleton and a backbone;" only corals, encrinites, crustaceans, and mollusks. "Vegetable appearances," he says, "do not appear among these British rocks; but there must have been a mass of vegetable life in the ancient sea, as no *fauna* can appear without a *flora* to uphold it." This last inference is of little immediate consequence; but I may remark, that it coincides with one which I ventured to make, prompted thereto by some of the recent papers of Mr. Murchison. We here see it sanctioned by a writer who is understood to be a distinguished investigator of the lowest fossiliferous beds. It is from no wish to amuse the reader, but merely as a pleading in behalf of several of the alleged geological mis-statements in my book, that I bring forward another distinguished reviewer of the *Vestiges of Creation*, (*North British Review*, No. 6,) taxing me with having been driven to make this very surmise as an escape from a difficulty! More than this: the North British reviewer is at odds with his Edinburgh brother, in bringing bones and teeth of fish into the first fossiliferous formation; grounding the statement upon authorities long antiquated, and contrasting with it, in a foot-note, my remark, "Neither fishes nor any higher vertebrata as yet roamed through the marine wilds." The fact is, that this last critic—understood to be a very eminent philosophical writer—was not aware, that of late the lower fossiliferous rocks have been divided into several distinct formations, in the lowest of which, it is fully admitted, there are no vertebrata. More than this still: a body called the Literary and Philosophical Society of Liverpool had brought before them (January, 1845) a set of letters which one of their members had drawn, with reference to my book, from several of the chief geologists of the day. We there find Mr. Lyell stating upon hearsay, that I represented fish beginning in the coal, and Mr. Murchison speaking of me as beginning with zoophytes and polypiaria alone; statements, I need hardly say, conveying the most erroneous impressions regarding the book. This, however, is not the immediate point. The two gentlemen here named will be allowed to stand in the very first rank as geologists. They are able men, of marvellous industry, and unimpeached zeal for science. These men, nevertheless, in the correspondence to which I am pointing, give entirely opposite views of the first fossiliferous formation. Mr. Murchison says, "No trace of a vertebrated animal has been found in the lower Silurian rocks." Mr. Lyell says, "The fact that, with the earliest type of organization, we meet with vertebrated animals, true fish, so far from being explained away since I affirmed it in my book, is confirmed and extended by fresh evidence." The very latest affirmation we have on this point from Mr. Murchison—an affirmation made after examining Silurian rocks in Russia, where they are presented in vast extent—contains these words: "The absence of even the lowest of the vertebrata in the inferior Silurian rocks,—*an absence which is total*, so far as can be inferred from the researches of geologists in all parts of the world,—gives them a true Protozoic character."* These extracts speak for themselves. The only thing calling for further remark, is the surprising circumstance of this correspondence having been brought before a learned society, as wholly and nothing else but a condemnation of the *Vestiges!* †

* Abstract of paper by Mr. Murchison, Report of British Association of 1844, page 54.

† See Examination of the theory contained in Vestiges of

A leading objection, with regard to the first fossiliferous formation (Lower Silurian) is, that it does not solely present animals of the lowest sub-kingdom, as corals and encrinites, but also examples of the two next higher sub-kingdoms, the articulata and mollusca, some of the latter being of the highest order, the cephalopods. The latter particular is what is chiefly insisted upon.

At the time when I wrote, it was understood that the highest orders of mollusca were not found in the first fossiliferous rocks. Professor Phillips, in 1839, (*Treatise on Geology*,) said, expressly, with regard to what was then called the Clay-slate and Grawacke system, "No gasteropods or cephalopods are as yet mentioned in these rocks in Britain; and we do not feel sufficiently acquainted with the geological age of the limestones of the Hartz, to introduce any of the fossils of that argillaceous range of mountains." So much as a justification of the view given of the Clay-slate fossils in my first edition. Since then, this formation, as it exists in England, has been found to contain gastero-

the Natural History of Creation. By the Rev. A. Hume. Liverpool, Whitby, 1845.

pods and cephalopods, though not of such high forms as afterwards appeared. I might here repeat what was remarked in the later editions of the *Vestiges*, "Even though the cephalopoda could be shown as pervading all the lowest fossiliferous strata, what more would the fact denote than that, in the first seas capable of containing any kind of animal life, the creative energy advanced it, in the space of one formation, (no one can tell how long a time this might be,) to the highest forms possible in that element, excepting such as were of vertebrate structure." I might add, that this was no great advance in comparison with the whole line of the animal kingdom, if we may take, as a criterion on this point, the analogous progress of an embryo of the highest animals, as the portion of that progress representing the organization of the invertebrated animals is *only the first month.* I might here also revert to the book for some views with respect to the space required for such a development. According to the plan of animated nature, to which I have made approaches in the later editions, we have not to account for the development of one long line, but of many comparatively short ones.

And, as I have also remarked, there is a rapidity of generation amongst the lower animals which may well suggest something like that "rush of life," which, if we were to judge from British strata alone, would seem to have taken place in the early seas. But there is no need for putting any of these speculative answers to the objection into requisition, while there is a preliminary question to be answered. Does the lowest band of the English lower Silurians indicate, beyond all question, the point of time at which animal life commenced upon our planet? Are we quite sure that cephalopoda were among the first of all earth's living creatures? Far from it. It has only been ascertained that certain comparatively small cephalopods are found as far down as any other animals of inferior organization at certain spots in Wales and Cumberland. When we remember that, in modern seas, certain kinds of such animals haunt special places suitable for their subsistence—that we may have crustacea and mollusks exclusively at one place, and radiata (as corals and zoophytes) at some other, not perhaps far distant, but different with respect to depth or some other circumstance—we can conceive that cephalopods may occur in the first fossil bands in the places which have been examined in England, and yet remains of inferior animals may be found by themselves on the same or a lower level in some as yet unexplored place not far off; so that a time-interval may there appear to allow for a progressive development. Such seems but a reasonably cautious surmise, when we are told by a high authority, that there are "detached Silurian districts in England, presenting particular changes and modifications, arising from difference of depth, and the variety of currents, and chemical combinations in the seas in which they were formed;" and that, "in consequence of this variety of physical condition, *there is a corresponding diversity in the traces of organic life in each situation.*" * What, however, places the matter beyond doubt is, that in North America, where the early stratified rocks are even more amply developed than with us, the highest invertebrated forms *do not appear at the first.* In the earliest ascertained fossiliferous strata, the Potsdam Sandstone, the only fossils are lingula (a brachiopodous mollusk) and fucoids. In the next, the Calciferous Sandrock, are fucoidal layers, encrinital

* Professor Phillips, British Association, 1845. Athenæum's Report.

beds, and the brachiopods, orthis, lingula, and bellerophon, together with orthocerata, *these being the first examples of the cephalopoda.* And in all these cases, the fossils are few and obscure; *they comprise no crustacea.* It is not till we ascend to a fourth fossiliferous series, Trenton Limestone, that fossils become abundant, or that trilobites appear. Perhaps even this is not the most decisively adverse view which could be derived from the American fossils, for lately there have been found, in the Green Mountains of Vermont, strata which, from their metamorphic character, are believed by some native geologists to be inferior and of course anterior to the Silurians, and these contain traces of fucoids and of vermiform bodies called Nereites, the last being a humble form of articulata. If this be true, it would at least add materially to the grounds for hesitation before pronouncing definitely, as the Edinburgh reviewer has done, on the commencement of fossiliferous strata and the nature of the first fossils. Here we must also remember, that in rocks of the elder continent anterior to the Silurians, there are limestones, held by many to be an indication of organic life at the places where they are found: the chemical experiments of Braconnot upon masses of these earlier rocks gave ammoniacal and combustible products, likewise indicative of the presence of organic matter: in the same sub-silurian region, " fragments, apparently organic, and resembling cases of infusoria," have been detected,* and in Bohemia actual fossils have been announced. Even dubious traces of life in sub-silurian rocks must be admitted to be of importance, when we consider that they have mostly been subjected to such a degree of heat as could not fail to obliterate organic memorials, seeing that it has even changed the texture of the rocks themselves. From what Mr. Lyell saw of the Silurian rocks in America, he finds himself called upon, in the most emphatic manner, to warn geologists against " *the hasty assumption, that in any of these sections we have positively arrived at the lowest stratum containing organic remains in the crust of the earth, or have discovered the first living beings which were imbedded in sediment.*"

" A geologist," he says, " whose observations had been confined to Switzerland, might imagine that the coal measures were the most ancient of the fossiliferous series. When he extended his investigations to Scotland, he might modify his

* Ansted's Geology, ii. 60.

views so far as to suppose that the Old Red Sandstone marked the beginning of the rocks charged with organic remains. He might, indeed, after a search of many years, admit that here and there some few and faint traces of fossils had been found in still older states, in Scotland; but he might naturally conclude, that all pre-existing fossiliferous formations must be very insignificant, since no pebbles containing organic remains have yet been detected in the conglomerates of the Old Red Sandstone. Great would be the surprise of such a theorist, when he learnt that in other parts of Europe, and still more particularly in North America, a great succession of antecedent strata had been discovered, capable, according to some of the ablest palæontologists, of constituting no less than three independent groups, each of them as important as the 'Old Red' or Devonian system, and as distinguishable from each other by their organic remains. Yet it would be consistent with methods of generalizing not uncommon on such subjects, if he still took for granted that in the lowest of these 'Transition' or Silurian rocks, he had at length arrived at the much-wished-for termination of the fossiliferous series, and that nature had begun her work precisely at the point where his retrospect happened then to terminate."*

It is exactly to such theorizers as the Edinburgh reviewer that this rebuke is applicable. When he asserts the contemporaneousness of the highest mollusks with the origin of organic life, he says —"We are describing phenomena that we have seen. We have spent years of active life among these ancient strata—looking for (and we might say longing for) some arrangement of the ancient fossils which might fall in with our preconceived notions of a natural ascending scale. But we looked in vain, and we were weak enough to bow to nature." The weakness consisted in looking only in one little portion of the earth, and believing it to be a criterion for all the rest. This writer seems yet to have to learn that knowledge is to be acquired by communication as well as examination. Were a philosopher (supposing there could be such a being) to limit his view of mankind to juvenile schools, he might with equal rationality deny that there is any such thing in the world as infants in arms. "We speak of what we have seen," he might say, "and, finding no specimens

* Travels in North Americ), ii. 131.

of humanity, under three feet high, we are weak enough to bow to nature and believe that babes are a mere fancy.

Even taking the English Lower Silurians as he and others would have them taken, it still appears that these rocks denote, generally, a low state of the animal kingdom. It is customary for those who take opposite views, to speak of the creatures of this period as high—"highly-organized crustacea and mollusca" is the usual phrase. Some, including the Upper Silurians in their view, tell us that the first formation presents examples of the whole of the great divisions, the fish being held as representing the vertebrata. Of course, this is only done through ignorance or for the purpose of deceiving. Where particulars are overlooked, it is still customary to speak of the earliest fauna as one of an elevated kind. When rigidly examined, it is not found to be so. In the first place it contains no fish. There were seas supporting crustacean and molluscan life, but *utterly devoid of a class of tenants who seem able to live in every example of that element which supports meaner creatures.* This single fact that only invertebrated animals now lived, is surely, in itself, a strong proof that, in the course of nature, *time* was necessary for the creation of the superior creatures. And, if so, it undoubtedly is a powerful evidence of such a theory of development as that which I have presented. If not so, let me hear any equally plausible reason for the great and amazing fact that seas were for numberless ages destitute of fish. I fix my opponents down to the consideration of this fact, so that no diversion respecting high mollusks shall avail them. But this is not all. The Silurian is an age, as were several subsequent ones, of only marine animals. It is now incontestable, from a few land-plants found in the Silurians of America, and a fern leaf in our own, that there was dry land; yet no trace of a land animal appears for ages afterwards. Moreover, though we have now a pretty full development of the first sub-kingdom, Radiata, we have but an imperfect one of the two next—namely, the Articulata and Mollusca. Not to speak of the utter absence of fresh-water and land mollusks, and of such land articulata as insects and spiders, we do not find any decapodous crustacea (crabs, &c.), though these could have lived wherever other mollusks and crustacea could. In fact, it is a scanty and most defective development of life; so much so, that Mr. Lyell calls it, par excellence, the Age of

Brachiopods, with reference to the by no means exalted bivalve shell-fish which forms its predominant class. Such being the actual state of the case, I must persist in describing even the fauna of this age, which we now know was not the first, as, generally speaking, such a humble exhibition of the animal kingdom as we might expect, upon the development theory, to find at an early stage of the history of organization.*

We now come to the *Upper Silurians*, where new species of invertebrated animals appear, besides a few obscure fishes. There is no appearance, according to the Edinburgh reviewer, of a transition from the former species to the present

* Objectors to the development theory have, in the eagerness of counter-theorizing, committed themselves on the subject of the Silurian fossils, in a way which they will yet feel to be extremely awkward. The *North British Review* we have seen placing even fishes in the first fossiliferous rocks, grounding this statement upon an authority which has been antiquated for fully eight years—a vast period in the history of geology. The *British Quarterly Review* is equally unfortunate. "The Author's theory," says this writer, "requires that these animals should be the lowest in the animal scale. But no argument can convert *a fish*, with its back-bone, and highly-developed nervous and muscular systems, into an animal of low organization." (!) The dogmatic allegations of the Edinburgh reviewer on this point are sufficiently exposed in the text. I have only further

—but does he know the signs by which such a transition could be detected? I am aware of none. He says the new species are sharply defined—that is, strongly distinct; and so they may be, without any prejudice to the transmutation theory—as far, at least, as I understand it. And here he remarks that there are the same difficulties in the way of his theory, "both in the grouping of each separate system, and in the passage from one system to another; and that is true, whatever part of the ascending geological series we choose to take between the lowest formations and the highest." As he does not state the nature of the difficulties, I cannot undertake to say what argument or what reconstruction of my system may be necessary to meet

to express surprise at finding Dr. Whewell participating in the mere ignorance of the first two of the above-mentioned journals. In the preface to a volume which he has recently published under the title of *Indications of the Creator*, he meets my arguments with a crude and incorrect view of the fossil history, commencing with this sentence—"Vertebrate animals do exist in the Silurian rocks, from which the asserted law [that of development] excludes them." The existence of a non-pisciferous formation had been unknown to him. Many of the objections made to the development theory, in obscurer quarters, rest on errors of a similar kind.

them. Till we are more clear, however, regarding the actual affinities of animals, I would suppose that any judgment as to difficulties in their grouping in geological formations, or succession in different formations, might well be given somewhat less dogmatically than they are by this writer.

The few fish-remains of the Upper Silurians may be associated with the ample development of this class in the next (*Devonian* or *Old Red Sandstone*) system. They belong to Agassiz's two orders of placoids (these by themselves in the Upper Silurians) and ganoids, the former of which are represented by our sharks and rays, the latter by the bony pike of America and the polypterus of the Nile. Such are the only fishes found till we come up to the chalk formation, when the now predominant orders of cycloids and ctenoids begin.* The Edinburgh reviewer

* The North British Review presents, as a strong objection, that, "several new ctenoids, which had been found only in the carboniferous system, have been discovered among the fishes brought by Mr. Murchison from the Old Red Sandstone of Russia. Resolved to make out his position, the author asserts," &c. This is an unlucky venture in opposition. The critic evidently meant it to have a very damaging effect, in consideration that the ctenoids are osseous fishes. The fact is, that the fishes brought

makes a strong point of the placoid and ganoid orders, as unfavourable to the progressive theory. "Taking into account," he says, "the brain, and the whole nervous, circulating, and generative system, the placoids stand at the highest point of a natural ascending scale, and the ganoids are also very highly organized." Of certain families of the first order, found in the Old Red Sandstone of Russia, he says, "Let the reader bear in mind that these fishes are among the very highest types of their class, and that we can reason upon them with certainty, because some of them belong to families now living in our seas." He instances a cestraceon — a high kind of placoid — recently found in the Wenlock limestone, a low portion of the Upper Silurians, and therefore near the beginning of fish. Some of the ganoids, also, of the Old Red Sandstone make an approach to a higher class—reptilia. Besides the usual row of fish-teeth, they have an inner range, in which we

home by Mr. Murchison are not of the ctenoid order, but belong to a placoidean family called Ctenodus. The mistakes made by this writer, in the geological part of his paper, are of a very grave kind, yet only such as many men of scientific eminence may be expected to make when they venture out of their own peculiar department, and rashly under-estimate the strength of the arguments to which they are opposed.

see the form of those organs among the sauria. It appears, in short, according to this writer, that the further back we go among the fishes, we find them possessed of the higher characters. Of the real character of all this hardy assertion I shall enable the reader to judge. The fishes of this early age, and of all other ages previous to the chalk, are for the most part cartilaginous. The cartilaginous fishes—*Chondropterigii* of Cuvier—are placed by that naturalist as a second series in his descending scale; being, however, he says, "in some measure *parallel to the first*." How far this is different from their being the highest types of the fish class, need not be largely insisted on. Linnæus, again, was so impressed by the low characters of many of this order, that he actually ranked them with the worms.* Some of the cartilaginous fishes, nevertheless, have certain peculiar features of organization, chiefly connected with reproduction, in which they excel other fish; but such features are partly partaken of by families in inferior sub-kingdoms, showing that they cannot

* Dr. Fletcher places the Chondropterigii lowest in a scale which takes as its criterion "an increase in the number and extent of the manifestations of life, or of the relations which an organized being bears to the external world."

truly be regarded as marks of grade in their own class. When we look to the great fundamental characters, particularly to the framework for the attachment of the muscles, what do we find?—why, that of these placoids—"the highest types of their class!"—it is barely possible to establish their being vertebrata at all, the back-bone having generally been too slight for preservation, although the vertebral columns of later fossil-fishes are as entire as those of any other animals. In many of them, traces can be observed of the muscles having been attached to the external plates, strikingly indicating their low grade as vertebrate animals. The Edinburgh reviewer's "highest types of their class" are, in reality, a separate series of that class,—generally inferior, taking the leading features of organization of structure as a criterion,—but, when details of organization are regarded, stretching further both downward and upward than the other series; so that, looking at one extremity, we are as much entitled to call them the lowest, as the reviewer, looking at another extremity, is to call them the highest of their class. Of the general inferiority, there can be no room for doubt. Their cartilaginous structure is, in the first place, analogous to the embryotic state of vertebrated animals in general.* The maxillary and inter-maxillary bones are in them rudimental. Their tails are finned on the under side only, an admitted feature of the salmon in an embryotic stage; and the mouth is placed on the under side of the head, also a mean and embryotic feature of structure. These characters are essential and important, whatever the Edinburgh reviewer may say to the contrary; they are the characters, which, above all, I am chiefly concerned in looking to, for they are features of embryotic progress, and embryotic progress is the grand key to the theory of development. I therefore throw back to my reviewer the charge that I have "clung to feeble analogies," and "kept out of view the broad and speaking facts of nature."

With regard to the alleged falsity of the crustacean character of some of these fishes, and the discredit of repeating the blunders and guesses made by the first observers, before any good evidence was before them, I can only say, that at the time when my book was written, geologists and inquirers into fossil ichthyology of the highest

* Cartilage, "in many animals, forms the entire structure, and in the early state of the human embryo it does the same."—*Carpenter's General Physiology*, p. 37.

character were writing, publicly and privately, of the cephalaspis and coccosteus, as apparently links between the crustacea and fish, the vertical mouth of the latter animal being particularly cited, as a feature indicating the intermediate character. In what the reviewer calls "the excellent work of our meritorious self-taught countryman," Mr. Hugh Miller, published in 1841, the apparently crustacean character of these fishes is repeatedly referred to.* Not having access at the time to the work of Agassiz, I deemed myself safe in trusting to the report of this industrious inquirer and ingenious writer, whose volume was then newly published.

The above argument relates to the general fact of the first fishes being placoidean. It is necessary, also, to meet the inquiry why there should

* Mr. Miller calls upon his readers to "mark the form of the cephalaspis, or buckler-head, a fish of the formation over that in which the remains of the trilobite most abound. He will find," he says, "the fish and crustacean are wonderfully alike: the fish is more elongated, but both possess the crescent-shaped head, and both the angular and apparently jointed body. They illustrate admirably how two distinct orders may meet. They exhibit the joints, if I may so speak, at which the plated fish is linked to the shelled crustacean. Now, the coccosteus is a stage further on; it is more unequivocally a fish; it is a cephalaspis, with a scale-covered tail attached to the angular body, and the horns of the crescent-shaped head cut off."—*Old Red Sandstone*, p. 54.

be no fossil remains indicating a transition from the lower animals to fish. The reviewer speaks of a recently discovered cestraceon below any other fish-beds in England. "Such," he exclaims, "are nature's first abortive efforts." "We entreat," he adds, "any good naturalist well to consider such facts as these, and tell us whether they do not utterly demolish every attempt to derive such organic structures from any inferior class of animal life found in the older strata?" Now, I cannot tell what good naturalists may say in answer to this appeal; but I feel, for my own part, that the facts in question—as far as they can be admitted to be so—have no such destructive effect.

In the first place, the cestraceon is only one of those cartilagines, the real character of which had just been explained. It is not the lowest of its order, but neither is it the highest. So far from this being the case, the respiration of the whole family (Selacii, *Cuv.*; Plagiostomi, *Desm.*) to which it belongs, and which also includes sharks, is performed in a manner which approximates these fishes to the worms and insects—namely, "by numerous vesicles called internal gills, the entrance to which is from their gullet, while the exit is in general by corresponding apertures on the sides of their neck;* other fishes having free gills, marking a higher organization. The subdivided form of the stomach—the absence of that concentration, which is, perhaps, the most emphatic mark of animal advancement—belongs to this family alone amongst fishes, as it does to the lowest families of several of the higher orders of the vertebrata. Thus, the cestraceon is, on many considerations, a low fish, though certainly possessing some traits of superior character, and not the lowest of its order. In the second place, I would protest against any inference unfavourable to the hypothesis of development being drawn from a discovery so new, so isolated, and in a branch of inquiry so extremely unsettled. At no time during the last ten years, have we had for a twelvemonth at once, stable views respecting the initiation of fishes. Lately—so lately that part of my book was written at the time—the lowest were understood to be some of a minute size, immediately over the Aymestry limestone, in the Upper Silurians.† Now, we have a cestraceon an-

* Fletcher's Physiology. Part 1, p. 20.

† "The minute and curious fishes in the uppermost bed of the Ludlow rock, are the *earliest precursors* of many singular ichthy-

nounced to us at a lower point in that formation. But how far it is likely that our information is to rest at this point the reader may judge, when he hears of M. Agassiz announcing, within the last few months, that, though acquainted with seventeen hundred species of fossil fishes he regards the history of the class as so far from complete, that the number of species successively entombed in the crust of the globe might be estimated at thirty thousand, without any chance of approaching the truth!* If such be the case, we may surely expect to hear of other fishes prior to or contemporary with the cestraceon, showing that, humble as that animal was, it is not to be regarded as the initial of its class.† But even although simpler

olites which succeed in that enormous formation, the Old Red Sandstone."—*Murchison's Address to the Geological Society*, February, 1842.

* Review of Professor Pictet's Traité Elémentaire de Palæontologie, translated in Jameson's Journal from the Bibliothèque Universelle de Genève, No. 112, 1845.

† Such shifts are of frequent occurrence in geology. Insects, formerly found first in the oolitic formation are now taken back to the carboniferous. Birds are now inferred from foot-tracks in the New Red Sandstone, their first place formerly being in the oolite. We have mammifers in the oolite, which a few years ago, were believed not to occur before the tertiary. None of these shifts, however, in the least interfere with the general fact of the advance from the lower to the higher classes of animals.

fishes be not found in lower or contemporary strata, this may only be owing, like the non-discovery of vegetation in the early rocks, to the unsuitableness of these fishes for being preserved. Supposing the inferior tribes, petromyzonidæ (lampreys) to have been then in existence, we should have no trace of them preserved, because of their osteological structure being slight, and their wanting those teeth and spines which form, after all, the chief memorials of the higher families of their own order.

One word more as to these fishes. The critic says (p. 38), it is shown to demonstration in the *Poissons Fossiles* of Agassiz, that "the sauroids, in their general osseous structure, and in the development of their nobler organs, run close upon the class of reptiles." There is no doubt that the sauroid fishes partake of reptilian characters, though, perhaps, in a more external and less important way than such writers as the Edinburgh reviewer suppose; but be it remembered, the sauroids are not the first fishes. There is not one of them in the Silurian formation, where placoideans appear to begin. Yet I do not, for this reason, suppose that the sauroids arose from placoideans. More probably, they are part of a distinct line of development, which had inferior forms in its first stages, also of too slight a structure to be preserved.

Following this reviewer into his discussion of the *Carboniferous System*, we find him commencing with a taunt, that there are now traces of land vegetation in earlier formations. This is, in reality, a point of no importance for the development theory. The question is, with what kind of plants did land vegetation begin? The anxiety of the reviewer to force a verdict in his favour is here strongly shown. "What," he says, "are these first fruits of nature's vegetable germs? Are they rude, ill-fashioned forms? Far otherwise. We find among them palms and tree-ferns, &c." In this passage, which substantially conveys the same information as my book, there is an evident design of inducing the belief, that the first land vegetation was of a high character. The rigid truth is, that though this was a "grand" in the sense of a luxuriant vegetation, it was composed, as far as positive evidence goes, almost wholly of plants which stand low in the scale of organization. The ascertained dicotyledons (plants having double-lobed seeds and an exterior growth) are extremely rare. On this point, I cannot do better than quote the laborious young Professor of King's College—"The plants which have hitherto been described [in the carboniferous formation], belong either to the acotyledonous class, as the ferns, or to the monocotyledons *and, on the whole, they constitute the simplest forms of vegetation;* but there have also been met with among coal plants, unquestionable evidences of dicotyledonous structure, and a genus has been formed under the name of Pinites, to include a number of specimens of fossil wood, &c."* To the undoubted evidence of Mr. Ansted, may be added that of his more eminent contemporary, Mr. Lyell, whose sense of the botanical character of this age is such that he emphatically calls it the *Age of Ferns.*† It is evident, then, taking the land-plants of this era as the first, that it is of a nature to harmonize with the development theory, for its chief forms are humble, and only a few are of higher grade, most of these, too, being of an intermediate character between the low and the high. I am reminded, however, in other quarters, of certain experiments of Dr. Lindley, showing that the plants chiefly found in the coal are

* Ansted's Geology. 1844.

† Travels in North America, ii. 52.

of the kinds which best resist decomposition in water; whence it is inferred that many trees of a high class may have existed at that time, but perished in the sea, while weaker vegetation survived. This evidence would be negative at the best; and it says as much for the non-preservation of mosses and other humble plants as for dicotyledons. It has also been remarked that, considering such facts as the disappearance of equisetum hyemale in water, a plant containing an unusual quantity of silex, "the proportion of fossil plants in each formation must depend on other circumstances besides their power of resisting decomposition."* "Too much importance has," in the opinion of the author of this observation, "been attached to Dr. Lindley's experiments."

The *British Quarterly Review* says—"The author admits there were dicotyledons among these plants, and does not see that, however few they may be, it entirely upsets the theory of progressive advance, especially in the absence of any proof as to whether they were created first or last." This proceeds, as do many similar objec-

* Mr. C. J. Bunbury, at the British Association, 1845; Athenæum's Report.

tions, upon the idea that a formation represents one point in time. A formation, in reality, represents many years, or rather ages. Such expressions as that simple and complex plants occur together in the carboniferous formation, or even (shall we say) in its first fossil bands, are vague expressions; perhaps, conveying an idea substantially false. There is no such precision in the ascertained relations of fossils to particular strata as to entitle any one to say that the simple and complex plants of this formation are rigidly contemporaneous. They may have followed each other within the space of half a century in a particular region, *and yet been preserved in but one stratum, or little group of strata.* The actual appearances of the carboniferous formation thus, perhaps, allow full time for a progressive advance in particular regions, from the fleshy luxuriant plants of the marsh and low sea-margin, to the robust tree of the more elevated regions. We must remember, too, that the vegetation of the carbonigenous era, even if we take it back to include the conifer said to have lately been found in the Old Red of Cromarty, or the fern leaf of the Silurians, was preceded by unequivocally simple plants in the fucoids. Starting with these, and finding the first great burst of land vegetation composed mainly of low cryptogamic and monocotyledonous plants,—finding, moreover, the exceptions chiefly of the intermediate character, and that the dicotyledons increase afterwards while the others decline,—we cannot well resist the conclusion, that we see the traces of a *progress* in the history of this kingdom of nature. It may be less clear than we could wish; but such light as we have certainly favours the development theory.

We now come to the *Magnesian Limestone* deposit, latterly called the *Permian System*. At this place, the Edinburgh reviewer introduces some general observations, which I hope he will yet acknowledge to be unjust, as I am sure the whole of his substantive charges are. "It may be true," he says, "that sea-weeds came first, but of this we have no proof." How a *good* geologist can have allowed himself to speak in this manner, even in eagerness to theorise against theory, I am quite at a loss to understand, for the positive facts of the occurrence of fucoids in the Lower Silurians, and of the very first traces of land vegetation in subsequent formations, are as palpable and undoubted as he himself acknowledges the precedence of fish by invertebrata to be; nor has any one ever pretended to expect that land vegetation would be found earlier than the marine. I have here ventured no conjecture of my own, but only spoken as all the geological books teach. "Of land plants," he continues, "we have not the shadow of proof that the simpler forms came into being before the more complex." The reader has just been told upon undoubted authority that, in the first great show of land vegetation, taking such positive evidence as we have, the simple forms are vastly more numerous than the complex. Finding that we have first ample marine vegetation, then a land vegetation in which the plants, with only a small exception, are cellular and cryptogamic, while of the exception a very small number are dicotyledonous, and a conspicuous group (the conifers) intermediate—I feel that I am entitled to say that positive evidence speaks for a precedence of high by simple forms; which is what I have done. "It is true," thus proceeds the reviewer, "that we see polypiaria, crinoidea, articulata, and mollusca; but it is not true that we meet with them in the order stated by our author." It is humiliating to have to answer an objection so mean. There is no statement that the animals came in this order. I have only put the *words* into this arrangement, in accordance with the custom now commonly followed of observing the ascending grades of the animal kingdom. With respect, then, to what follows—"The sentence on which we here comment contains three distinct propositions, and all three are false to nature, and no better than a dream,"—I believe I may safely leave the reader to say which party is the falsifier and the dreamer. He goes on in the same strain—"It is true that the next step gives us fishes; but it is not true that the earliest fishes link on to the radiata: this is a grand and at the present day an unpardonable blunder." This is another dream of the reviewer, for certainly such an affinity was not suggested in any edition of the *Vestiges* hitherto published. In the first four editions, which alone were under his notice, no passage except from the articulata was even hinted at. So much as a proof of the reviewer's recklessness in making charges; there is no need, however, to affirm, with him, that a connexion between certain high radiates and some of the lowest fishes does not exist. I venture to predict that affinities of an equally startling nature will yet be made familiar to naturalists. Meanwhile, it is enough to show that this confident critic has raised an accusation for which he has not a shadow of ground.

Taking up the special fossils of the Permian system, he says, "The earliest reptiles are not of such a structure as to link themselves, on a natural scale, to the noble sauroids of the preceding carboniferous epoch." They are not the marine saurians, or fish lizards (ichthyosauri) which occur in a higher formation, but lacertilians, or animals of lizard-like character. Now what first strikes me here is the extraordinary narrowness of a mind which sees nothing indicative of natural procedure, no hint towards great generalizations, in the simple fact of reptiles following upon fish in this grand march of life through the morning time of the world. He knows that, in every classification of the animal kingdom, reptiles rank next above fish, that in some living families there is such a convention and intermixture of both characters, that naturalists cannot agree to which class they should be assigned. He actually sees, in a general view of the earlier reptiliferous formations, animals combining the fish and reptile in the most unequivocal manner. Despising, however, the great fact which shines through these obscurities, this person, and I am sorry to add, geologists generally, can only fasten upon such particulars as may be made out to be difficulties in the way of generalization. Passing to the particulars, a few land lacertilians come first, whereas the first, according to my hypothesis, ought to be marine forms, and linked to fish. He says of this difficulty, that I have stated it feebly. Perhaps it would have been well for his own credit that he had stated it somewhat less confidently; for before his sheets had seen the light, a prospect had arisen of his affirmations on this point being thoroughly falsified. In *Silliman's Journal*, for April 1845, is an account of sandstone surfaces pretty far down in the *Carboniferous formation* of Pensylvania, marked with the vestiges of terrestrial animals. Setting aside in the meantime one class of these markings, which are said to indicate wading birds, we have a variety of others plainly denoting REPTILES. In one group, the foot consists of a ball, with five toes radiating from it in front. In another, the impression resembles that made by a coarse human hand, with the rudiment of a sixth toe at the outside. The reptilian families indicated by these foot-marks have not yet been pronounced upon, as far as I am aware; but from the extreme resemblance of some of them to the vestiges of the labyrinthidon, there can hardly be a doubt that some of the order batrachia are amongst them. If they prove wholly batrachian, as is not unlikely, for we have living families with feet resembling the first group of vestiges, or even if only a portion of them be certified as of this order, where will be the lacertilians, and where the confident counter-assertions of the Edinburgh reviewer? The batrachia he has himself allowed to be a low order of reptiles (p. 51.) They are so considered by all naturalists. Might I not here, then, take my stand upon the fact of animals, the lowest apparently of the reptile order, being now found at the earliest point of time? I might unquestionably do so with a decided immediate advantage to my hypothesis. It would in a great measure neutralise the whole of the objections of the reviewer with regard to the chronology of the reptiles. But I am, whatever he may think of me, willing to read the book of nature aright. I receive the fact as one liable any day to receive a new aspect from fresh discoveries. In as far as it is so, it only teaches that we are not to be too confident in drawing inferences either for or against the theory of development from the particular succession in which the *orders* of the reptilia occur in those early strata *where their remains and vestiges are few.* In as far as it may be taken as a positive fact, I only claim a modified benefit from it, because the view which I take of the affinities and connexions of the animal kingdom (and by analogy of the vegetable kingdom also) makes it a matter of less consequence than would be generally supposed, which *order* of any class appears first in the stone record, though still perhaps a matter of *some* consequence.

This view suggests that development has not proceeded, as is usually assumed, upon a single line which would require all the orders of animals to be placed one after another, but *in a plurality of lines in which the orders, and even minuter subdivisions, of each class, are ranged side by side.* It also suggests that the development of these various lines has proceeded independently in various regions of the earth, so as to lead to forms not everywhere so like as to fall within our ideas of specific character, but generally, or in some more vague degree, alike. The progress of the lines becomes clearest when we advance into the vertebrate sub-kingdom. We can there trace several of them with tolerable distinctness, as they singly pass through the four classes of Fishes, Reptiles, Birds, and Mammals; the Birds, however, being a branch in some part derived equally with the reptiles from fishes, and thus leaving some of the mammal order in immediate connexion with the reptiles. The lines or *stirpes* have all of them peculiar characters which persist throughout the various grades of being passed through, one presenting carnivorous, another gentle and innocent animals, and so on. We have, therefore, in the animal kingdom, not one long range of affinities, but a number of short series, in each of which a certain general character is observable, though not always to the exclusion of the organic peculiarities of families in neighbouring lines, especially in the class of reptiles.

According to this view, the matrix of organic life is, speaking generally, the sea. Fluid, required for all embryotic conditions, is also necessary to the origination of the various stirpes of both kingdoms. The whole of the lowest animal sub-kingdom (Radiata) is aquatic; so are nearly all the Mollusca and a very large proportion of the Articulata. In the Vertebrata, the lowest class also is wholly aquatic. The arrangement appears to be this—the basis of each line is a series of marine forms; the remainder consists of a series designed to breathe the atmosphere and live upon land, these being all of improved organization. The classification which this system implies may be said to be transverse to all ordinary classifications. The invertebrate, ichthyic, reptilian, ornithic, and mammalian characters are horizontal grades, through which the lines pass, and where they send off branches; not separate and independent divisions. In any of these branches where we have a clear knowledge of the various forms, it is possible to trace the affinities, in conjunction with an improved organization, through genera which are adapted to a partially marine life, to a residence in the mouths of rivers, or on shores and muddy shallows, then through genera which are, in succession, appropriate to marshes, jungles, dry elevated plains, and mountains. And it is this series of external conditions and adaptations which has caused that system of analogies between various families of animals which has of late attracted attention. But the immediate cause of the development of each line through its various general grades of being is to be sought in an internal impulse, the nature of which is unknown to us, but which resembles the equally mysterious impulse by which an individual embryo is passed through its succession of grades until ushered into mature existence. Geology shows us each line taking a long series of ages to advance from its humble invertebrate effluents to its highest mammalian forms; and this I have ventured to call "the universal gestation of Nature."

The traces of this order of the animal kingdom have been seen in all ages of science. Every zoologist acknowledges the gradations and affinities which appear amongst animals. Prompted by what so palpably meets observation, many have tried to range the various orders or families in one line, or (to use the favourite phrase) chain of being; but they have always failed, which is not to be wondered at. One cause why zoologists have not up to this time thought of trying any different arrangement, is the confusion arising from the prevalence amongst many families of parallelisms of structure, which have been regarded as affinities, when in reality they are only identical characters demanded by common conditions, or resulting from equality of grade in the scale. True affinities—and these are the affinities of genealogy—are not to be looked for horizontally amongst orders, but vertically, from an order in one class to the corresponding order in the class next higher. Generally, the first and lowest forms of the orders in a class are marine, and often these are of comparatively large size. We usually see in them a vestige of the essential characters of the class next below. Thus, the perennibranchiate batrachia in their order, and the ichthyosauri in the series of crocodilia, exhibit an affinity to fish. The cetacea and phocidæ, which I regard as the immediate basis of the pachydermata, carnivora, and other orders of terrestrial mammals, ought, according to this view, to show an alliance to the reptiles; and such a connexion does exist between the cetacea and certain marine sauria; but from the general extinction of the marine reptiles, the linking of the mammals to that lower class is less clearly seen than might be wished. It must be kept in view that only an outline of the progress of the animal kingdom is here designed. Exceptions as to the course which development has taken appear to be by no means few; leading to the idea that the grades of organization are not determinate in this respect, but may be reached by steps of unequal length. Thus, for example, the marsupials appear very clearly a development from certain birds; probably the rodent, edentate, and insectivorous orders are derived through the same channel. In short, the progress of animality in the different stirpes has been attended by peculiarities which evidently affix peculiar characters to each, and make the idea of a difference in *time* not only probable, but unavoidable.

Regarding the animal kingdom simply as a combination of independent stirpes, each with its distinct affinities, the theory of transmutation puts on a totally new aspect; so truly is this the case, that transmutation is hardly any longer a term appropriate to the idea. The difficulty of supposing such changes as that from the rodent to the ruminant, or the carnivorous animal to the quadrumane, vanishes, leaving only *transitions from one form to another of a series generally similar*—from the otary, for instance, to the otter, from certain phocæ to the bear, and so on. There is a unity in all instances in the moral as well as physical characters of the various members of one stirps; we only see it advancing from low to high characters, just as we see the fœtus of a high animal passing through various inferior stages before it reach its proper mature character. The lines, moreover, being independent of each other, and not quite uniform as to the stages of animality through which they pass, it follows that, unless we knew of some law governing their different gestative periods, we are not entitled to look for the first occurrence of their various ichthyic, reptilian, and mammalian sections, in any order as regards each other, even though we could be sure (which we are not) that we are surveying a geographical region where they all started fair in the race of progressive organization. Hence it is that, though the batrachia are usually placed by zoologists at the bottom of the list of reptilian orders, I attach little importance to their vestiges being now found so low. All that I think we can expect is, that, in a particular area where we have reason to believe that the lines have started abreast, they should all reach their various grades nearly about one time, or what may be considered as one time compared with the whole extent of geological chronology. And such appears to be pretty much the case in those regions which geologists have explored.

The Edinburgh reviewer will observe that this view of the animal kingdom leaves much of his opposition in a very awkward predicament. He has everywhere assumed that the genealogy of the orders of each class was supposed to be *en suite*, which it certainly never was in my book. In the early editions, the course of the supposed development was spoken of with diffidence,* because I had not then seen or conceived any arrangement of the animal kingdom which answered to that hypothesis, although I thought proper to attempt to show that the quinarian and circular classification, then or recently in vogue, did not necessarily militate against it. In the third edition, the present view was first hinted at; and in the fourth it was sketched, though with liability to correction; thus anticipating by some months the publication of the criticism now under notice. It is hardly necessary to remark, that in all criticism, the actual subject criticized must be brought forward for comment, and nothing else; otherwise the commentaries

* " . . it does not appear that this gradation passes along one line, on which every animal form can be, as it were, strung; there may be branching or double lines at some places," &c.—*Vestiges*, 1*st ed. p.* 191.

become of no imaginable use but to obscure true judgment. Now the Edinburgh reviewer has presented his subject, in this instance, in lineaments entirely of his own imagining, and directly in contradiction to those which belong to it. He had no title to assume any plan of development, and to represent his victory over *that* as a triumph over the hypothesis of his author. In such conduct, he has thoroughly vitiated the whole fabric of his criticism, and left it, in reality, no pretension to remain for a moment in court. My immediate object, however, is not to take such exceptions, but to show how the ascertained facts of a limited portion of the field of nature may be reconciled with that conception to which a view of what appears over the whole field may lead an honest inquirer.

If the hypothesis of a plurality of genetic lines be admitted, we are not of course to ask which *order* of reptiles, or of any other *class*, first existed, (such being the language of the old classification;) but, having first settled the whole affinities of the animal kingdom on the new plan, we are to inquire if the geological presentment of the *families* was accordant with the scheme, allowing for the negative nature of much of the geological evidence of this kind. Now, in the first place, the affinities of the animal kingdom are only in part made out; in the second, geological evidence is only partial. We are clearly, therefore, not to expect in nature's museum a full exhibition of any one entire stirps, as it may be supposed to have passed through its successive stages up to our time. All that we can expect is a succession of fossils marking out portions of what we may suppose likely yet to be established as lines of animal descent. Blanks, and large ones too, must be allowed for; possible errors as to the animal pedigrees must be contemplated. But, if we have any ground for generalising in a particular direction, as I think there is in this case, we may be held as called upon not to conclude hastily and rashly on the unfavourable side, but to look and consider patiently, and to suspend judgment wherever the adverse evidence may appear to be of a nature likely to be reversed. Let us now see how all this applies to the conduct of the Edinburgh reviewer, with regard to the early reptilian fossils. The formations where these occur have only been examined in such a degree, that they are almost every year giving forth new responses: for example, the existence of birds at this era was not dreamt of ten years ago; the existence of tortoises in the time of the New Red Sandstone was equally unknown only two or three years earlier. It is a still less time since the labyrinthidonts of the Keuper of Germany were discovered; and we have just seen that the unqualified affirmations of the Edinburgh reviewer, as to the oldest reptiles, were overturned by intelligence from America, before his sheets had seen the light. When these things are considered, we must see the objections of the reviewer to be extremely rash. It might be allowed that the earliest known lacertilia are not of strictly marine forms or allied to fish; it might equally be admitted of the first batrachians, that "their near affinities are not with fishes," as this writer takes it upon him to say. Yet we should still see the absurdity of affirming that either these batrachia or lacertilia were the first created of their respective orders, seeing that their relics were so few and the discovery of these so accidental, that we might look for new and superseding facts every day.*

* It is necessary to guard against a supposition that I undervalue such isolated relics, as inferring the positive fact of the existence of particular orders of animals at particular times. For this purpose, the smallest fragment betraying the character of the organization is often sufficient. What is really meant is, that,

But, as the case actually stands, is this line of defence more than hypothetically necessary? The lacertilia of the magnesian limestone, and these labyrinthidonts of the Trias, (perhaps also of the carboniferous formation,) are they so far removed from fish characters as the reviewer would make them? Let any naturalist who has ever studied the transmutation of the individual batrachian, passing in a few weeks from the branchiated fish to the lunged and limbed frog or newt, its circulatory and alimentary system entirely changed, say if the labyrinthidon may not be the very first step from some ichthyic form. What though the proportions of the head remind Mr. Owen of the sauria, and remove the animal, as he thinks, above the present batrachian type! Against any such inferences we have the positive fact, in the organization of this batrachian, of a biconcave form of the vertebræ, *the form peculiar to fishes*,—arguing, by Mr. Owen's own acknowledgment, aquatic if not marine habits,—also a decidedly piscine character in the arrangement and even microscopic structure

when we find a few outlying relics belonging to a class which does not appear in any force till afterwards, we cannot be sure that we have acquired the means of forming a distinct idea of *the time of the origin of that class*, or the orders with which the class started, as further discoveries on these points may be looked for.

of the teeth, together with that position of the breathing apertures near the end of the snout which we see in crocodiles, for the purpose of allowing them to drag their prey under water without ceasing to respire. With regard to the lacertilia, we have this same fish-like biconcave form of the vertebræ, and the same fish-like arrangement of the teeth, equally arguing that alliance to the lower vertebrate class which it is the pleasure of this hardy critic to deny,—the biconcave structure of the reptiles, showing, as Mr. Owen himself owns, that these animals, which the Edinburgh reviewer deems so utterly separated from fish, had probably "*a more aquatic, if not marine theatre of life*,"* than was assigned to their successors. In subsequent and present reptiles, this form is superseded by the ball and socket, or concavo-convex form; but it is remarkable that, in the embryo state, the frog and crocodile (if not others) exhibit the double hollow form still resembling in this respect the mature animal of the secondary rocks. Such is the actual character of reptiles which our critic would set up as high: he has, after this, only to speak of the annelid as

* On the Reptilian Fossils of South Africa. Geological Transactions, Feb. 1845.

above the butterfly, or the proteus as superior to the land salamander, to establish his character as a naturalist. Need I say that these Permian reptiles are, in reality, by these facts degraded to a place in proximity with fishes?

So much for the batrachia and lacertilia. When we come to the great saurian line in the Muschelkalk, Lias, Oolite, and Wealden, we have a case which cannot be disputed, for here the marine character of the earliest of the series, and their intermediateness between fish and true crocodiles are admitted by all. The first remove from the fish is the ichthyosaur, its name declaring the convention of class characters for which it is remarkable. With piscine body and tail, and fins advanced into a paddle form, it has a true crocodilian head. In the pliosaur, which is later in appearing, we have a stage of advance to the true sauria, which come forward in the oolite, in the forms of teleosaurus, steneosaurus, &c. Afterwards, chiefly in the Wealden, we have the dinosauria, which betray an approach to the mammalian type. Another oolite saurian, the cetiosaur, exhibits in the form of the vertebræ a verging towards the cetaceous mammalia. Here there is the most perfect and even striking harmony with the theory of a progressive development. Below these formations, fish: then, low in these formations, fish saurians; above them, true and complete saurians; finally, higher still, saurians advancing to a more elevated grade of animality; and where do these more elevated types occur? In the next formation, passing over one which hardly represents any but deep-sea life. Nay, cetaceous relics have been found before we leave the strata so remarkable for the saurians. Thus, it appears that the whole of this chapter of palæontology, when read by a light from nature, and not from man's capricious humour, so far from being opposed to the natural genesis of animals, gives it support. Men, however, and of lively parts too, might go on for an age misreading such palpable facts, if they be determined against putting them into the collocation in which a sense can be made of them, just as we might puzzle for ever over a Latin or Greek sentence, if obstinately resolved against making English out of it except in its original construction.

After presenting the case of the reptilian fossils of the secondary formation in this way, I feel it hardly necessary to track the Edinburgh reviewer through all his particular objections. They are a mass of confusion, resulting from erroneous assumptions on his own part respecting the development theory, as that the *orders* of animals are all to be affiliated to each other, and every parental form held as extinguished by the fact of transmutation (the latter being a peculiarly gratuitous supposition—see p. 50 of the Review); together with equally rash and unjustified conclusions regarding the earliest forms of the reptilian orders, all mixed up in the way that promised to tell most effectually in favour of his own opinion, and with a disregard of every thing that pointed in the opposite direction. The great unquestioned facts of a succession of birds and mammals to the fishes and reptiles, these being also the next higher classes in the scale of the naturalist, tell nothing to this writer, as the succession of the reptiles to the fishes told nothing before. From the slight remarks with which he passes over these facts, an unlearned reader would hardly suppose that they were of the least significance, while, in reality, they are of the greatest. It is much the same as if an historian were to sink all such events as changes of dynasties, and fix attention upon the displacement of under-secretaries of state. And what makes this conduct the more marked is, that the minor facts upon which he fastens for the purpose of supporting his own theory, are mostly presented to us in circumstances which show their uncertainty and the likelihood of their being superseded.

For example, the earliest traces of birds do not indicate marine forms, which, according to my general views, ought, he says, to be the case. Instead of natatorial birds, they are waders and runners. Let the reader judge of the character of this objection, when he learns the real circumstances of the case. The traces of birds here spoken of are merely a few foot-prints found upon certain rock surfaces in America. Not a bone of these animals has been found in this early period. It must therefore be inferred, either that the circumstances were not favourable for the entombment of the bodies of these birds, or that our researches in the strata formed at the time when they lived have been insufficient to discover them. If such be the case with birds which lived upon shores,—places where, as we learn from the nature of the strata, accumulations of sand and mud were constantly taking place,—it is of course not to be expected that any remains of natatorial birds should be found, animals mostly living far out at sea. To put the case in its strongest form—foot-prints on shores being the record of the birds of this era, we are not to expect any trace of such birds as, generally speaking, are not in the way of making foot-prints on shores. I might go further than this, and point out that certain natatorial genera have feet not to be distinguished from those of waders, so that certain of these foot-prints may be those of natatorial species after all; but I feel it to be my best duty in the case, only to deny that we are in circumstances to say that waders and runners were the first created birds. Mr. Lyell, who stands as high as this or any other writer on geology, says, with regard to these very ornithichnites, as they are called—"This sandstone is of much higher antiquity than any formation in which fossil bones or any other indications of birds have been detected in Europe. Still we have no ground for inferring from such facts, that the feathered tribe made its first appearance in the western hemisphere at this period. *It is too common a fallacy to fix the era of the first creation of each tribe of plants or animals, and even of animated beings in general, at the precise point where our present retrospective knowledge happens to* *stop.*"* What now gives force to this observation is, the recent discovery of a new set of bird foot-prints —said to be of waders only—in the carboniferous formation of Pensylvania. The emergence of such a fact in the midst of the reviewer's speculations on the foot-prints of the New Red Sandstone, forms a most emphatic commentary on all decisive inferences where the facts are obviously casual and isolated.

Of a somewhat different character are the reviewer's remarks on the first relics of mammalia —the few bones of cetacea from the Lower Oolite and of marsupials from the Stonesfield Slate. Here the very first mammal family is undoubtedly marine; and, if it were to receive equal consideration with the grallatorial foot-prints, he ought certainly to admit that it favours the development theory. But he escapes from this claim by a mode of his own. He has not *seen* these relics! The American foot-prints were good evidence, without being seen; but a fact which makes *against* his theory requires personal inspection, even though it may come forward with the authority of Baron Cuvier.† He is more at ease with

* Travels in North America, I. 255.

† "There is in the Oxford Museum an ulna from the Great

the marsupials, which are of course unequivocally land animals. I have only here to refer to the fourth edition of my book,—published two months before the appearance of the review, and while I was unrecking of any great objection being grounded on this point—where it is suggested that the peculiar organization of the marsupials points to their having been derived through a different medium from other mammals. The critic, eager to let nothing escape, tells us that there are other land mammals lower in organic type than the marsupials. One answer to this objection might be found in an explanation of my views respecting the ornithic descent of these animals; but I am unwilling to pause upon such an inferior matter, and will therefore meet him with the question, if any other mammals show that lowly grade of organization which is marked by the absence of a placenta? "There are no other organic types," he says, "to which they [the marsupials] offer the shadow of a near affinity. They are therefore in direct antagonism with the scheme of regular development." To this it may

Oolite of Enstone, near Woodstock, Oxon, which was *examined by Cuvier and pronounced to be cetaceous;* and also a portion of a very large rib, apparently of a whale, from the same locality." *Buckland's Bridgewater Treatise*, I. 115, *note.*

be replied, that the affinity of the marsupials to the oviparous vertebrata is admitted by every naturalist, being shown in the small size of the brain and consequent exposure of the cerebellum, the absence of the septum lucidum and corpus callosum in the brain, and various other traits. Professor Agardh says—"The marsupials are mammalia which approach very nearly to birds; the monotremata, in particular, almost coincide with them."* Professor Rymer Jones, of King's College, whose testimony on such a point will be admitted by the reviewer, speaks of the marsupials as "connecting links between the oviparous and placental vertebrata." Striking traits of their affinity to birds are shown, he says, in the structure of the ear and of the reproductive organs.† In reality, the whole figure of the cursorial bird, the small head upon the long neck, the extreme length of the hinder limbs, and the imperfect development of the fore extremities, as well as the tendency of the feathers to a hair-like character, speak irresistibly for its approach to certain marsupials. The ornithorhynchus is as clearly an advance from the natatorial bird towards the rodent form, the latter

* Allman Wext Biologi—*apud* Charlesworth's Magazine, July, 1839.

† General View of the Structure of the Animal Kingdom.

being an order whose osteological structure is allowed by every naturalist to be bird-like. New and curious illustrations of the connexion between the birds and the implacental mammalia are constantly appearing. We lately heard of a bird which has a pouch for its young like the kangaroo,* and Mayer has discovered in the female emeu a purse form of certain organs, indicating an approach to the marsupial in that part of structure which is the most distinctive in the case.† It would appear that the reviewer is simply ignorant of this department of natural history, and, with the self-esteem which often attends upon ignorance, he has somewhat unluckily ventured to give a positive contradiction to that which is incontestably true.

The reviewer at length comes to the organic phenomena of the Tertiary system. "On the theory of development," says he, "'the stages of advance are in all cases very small—from species to species,' and the phenomena, 'as shown in the pages of geology, are always of a simple and modest character.' Let us test these assumptions by one single step, from the chalk to the London

* Magazine of Natural History.

† Reports of Ray Society, I.

clay, or any other tertiary deposit. Among the millions of organic forms, from corals up to mammals, we find hardly so much as one single secondary species." The exceptions in reality are, the infusoria of the chalk, and "two or three secondary species," which are said to "straggle into the tertiary system." "Organic nature," he says, "is once more on a new pattern—plants as well as animals are changed. It might seem as if we had been transported to a new planet; for neither in the arrangement of the genera and species, nor in their affinities with the types of an older world, is there the shadow of any approach to a regular plan of organic development." Now the almost total break in the organic creation here insisted upon, occurs in the interval between the extensive deposits of the secondary formation, and the comparatively isolated deposits of the tertiary. It is an interval which the lithological arrangements clearly indicate to have been longer than any of those between the other formations, during which minor changes of organic creation had taken place. It is simply, then, a period not represented by strata or by fossils; while it elapsed, the continual advance of the organic world proceeded to a point at which nearly all the old species had died out or been changed. There was nothing more in the "step" of our reviewer than this. Such is the geological doctrine.

"Is the present creation of life," says Professor Phillips, "a continuation of the previous ones; a term of the same long series of communicated being? I answer, yes."* "There is no break," he says, "in the vast chain of organic development till we reach the existing order of things." The reader will further be able to judge of the candour of the reviewer respecting the zoology of the tertiary, when he is reminded that it shows exactly those new portions of the animal kingdom which might have been expected, according to the theory of development. Heretofore, we have only few and faint traces of mammalia; but now they are added in abundance, mammalia being the crowning class of the vertebrated form. As far as *class*, therefore, is concerned, it is incontestably a "regular plan of organic development." But this is not all. We have seen the reptile forms of the secondary approaching the cetacean character; and now there is an abundance of the *aquatic mammalia*, as well as of those *land pachyderms* which are

* He adds—"But not as the offspring is a continuation of the parent."

universally classed with some of the forms of that order, these being the only suite of creatures which my ideas of development would lead me to expect at this place. Here I must meet the reviewer on a special ground. He admits the dinosaurs to have been the nearest approach to mammals; but "they died away," he says, ("if we are to trust to geology,) ages before the end of the chalk." These mammals have, therefore, "no zoological base to rest upon." That is, there is no connexion between them and any such animals as the dinosaurs, because there is an interval in the cretaceous formation which gives neither these forms nor any intermediate. Now, the fact is admitted by Professor Ansted, that the cretaceous system appears to have been "formed, for the most part, by deposits in deep water, and a considerable portion of it *not far from the zero of animal life*."* And this he states with a particular reference to the results of Professor Edward Forbes's researches in the Egean sea. We therefore have a satisfactory explanation of the non-appearance of forms intermediate to the reptiles and mammals in the chalk, without being driven to suppose, with our reviewer, that the latter were a creation

* Ansted's Geology, I. 502.

de novo of animal life. But no such fact as this did it suit our reviewer to state.

"Carnivora," he proceeds to say, "are as old as pachyderms. As far, at least, as we have any evidence bearing on the question, and bimana (monkeys) are found in this division—thus contradicting and stultifying the upper end of our author's grand creative scale." There is here, in reality, no stultification except in the critic's own mind. It was not my scale which he refers to, but Dr. Fletcher's; adopted into my book, not as a plan of the actual process of development, but as a general indication of the comparative organization of the animal orders. I do not consider the assumed contemporaneousness of the carnivora and monkeys (which the reviewer erroneously calls bimana) as at all contradictory of a true development theory, for I regard them all as distinct lines of development, which might well advance to a certain stage, (namely, that of the terrestrial mammal) about the same time. I am not, however, entitled to blame the reviewer for this objection, as the idea of a development in a plurality of lines must be new to him.

"As we ascend," he says, "towards the middle divisions of the [tertiary] series, there is a development of nature's kingdom, nearer and nearer to living types. But it is not a development after our author's scheme. It follows the law of the rise, progress, and decline of the families of the older world, already pointed out. We have no confusion of genera and species, and no shades of structure to make dim their outlines." Now there is here an acknowledgment, in which all geologists accord, of a constant gradual approach to living types. Is not this, in itself, a fact speaking strongly for some simply natural procedure in the origin of the present tribes? A change goes on from one set of forms to another, in the same way as one human generation is changed for another—namely, by the withdrawal of some and the addition of others, until at length the whole *personnel* of one age is superseded by that of another. The removal of old species is the result, by our critic's own showing, of law; and laws for the extinction of species are in operation at the present day. Can we well suppose the rise of the new species to be a phenomenon of an essentially different character? for here is the whole question at issue. I say, no—any ideas I have ever acquired of philosophy, as an expression of our ascertainment of the order of nature or providence, forbid me to form such a conclusion. A "confusion of genera or species" is not to be presumed; there is no need for a shading of structure to make dim their outlines. I suggest, that a line of organization, analogous to the progress of the embryo of an elevated species, had passed in the course of time through its appointed stages of development, each of which is a small advance upon the preceding, and the type of a form thenceforth to continue permanent. Each line stands apart. It may show shadings in a vertical direction, as between its reptilian and its mammal forms, but no true affinities connecting horizontally with the members of other lines. Our critic is here, therefore, completely at fault. I meet him again, however, on special grounds. Many of the animals of the tertiary period are of large bulk. We have not only huge examples of the pachyderm order, in which there are still existing many bulky species, but we have creatures equally vast in proportion belonging to the rodent, the edentate, and other orders. These huge mammals are, indeed, the signal forms of this period, the forms by which the whole tertiary system is most distinguished. Now, if we take the living pachyderm order, we shall find that the largest species are of the lowest organization.

For example, the elephant, with its short metatarsus, is a low form compared with the horse, in which the heel is raised so much above the ground. This is a progress of characters which could be shown in many other families. It is a progress which may be generally described as passing from the phocal form of the hind extremities, through the plantigrade, and ascending to its ultimatum in the digitigrade. Now this progress is coincident with the distribution of the various lines of animals in physical geography, for while the first are marine, the second are generally found in connexion with shores, rivers, and low grounds, and the last (always the smallest) with the more varied surface of the interior. When we find, then, animals of the *second kind* most conspicuous in this period, we have actual phenomena remarkably in accordance with the scheme of development. We look in, as it were, upon the world, or at least, its chief zoological province, at the time when the lines had attained to the terrestrial mammal forms fitted for fluviatile and jungle life, and ere from these had yet sprung the whole of the smaller but more highly organized denizens of nature's common.

Our critic, having now run over the whole series

of fossils, summons Cuvier, Agassiz, and Owen to express their opinions against the theory of development. The first "again and again affirms that the extinct fossil species were not produced by any continued natural organic law from other species." His French opponents tried, according to the reviewer, to overturn his conclusion by experiments in cross breeding and the ransacking of ancient tombs. And they talked contemptuously of *la clôture du siècle de Cuvier*; for which they fall under a reference to the fable of the ass and the dead lion. Now, I disclaim all responsibility for the experiments and language of the French theorisers on this subject. But, while I respect Cuvier, I must not concede too much even to his opinion. He was, after all, but a man, with the common liability to prejudices. I would, with all due reverence for the illustrious Baron, remind my reviewer of an opinion which the former expressed in 1826, that a deluge had occurred about six thousand years ago, which broke down and made to disappear the countries which had before been inhabited by men, and the species of animals with which we are best acquainted. Ten years after this belief was expressed by Cuvier, I find Dr. Buckland quietly withdrawing his adherence to it in the Bridgewater Treatise. At this moment it is not supported by a single geologist of the least repute. May not, then, the Baron Cuvier be wrong also in his opinion regarding the development of species? So much, I trust, may be said without any disparagement to the author of the *Regne Animal.* The fact is, that the erroneous and imperfect ideas of great men often become an annoyance, from no fault on their part, but only because the weak and narrow-minded are so apt, afterwards, to seize upon such ideas, and brandish them in the faces of advancing truths. For M. Agassiz I likewise entertain great respect; but it happens that his liability to error is equally well established. The doctrines which he persisted for years in maintaining with respect to the constitution and movement of glaciers, are now all but deserted for the more accurate and philosophical deductions of Professor James Forbes. I may, therefore, receive the intelligence which the Neufchatel philosopher brings me regarding the fossil fish, but be cautious in accepting as an infallible dictum what he is pleased to say on the comparatively profound doctrine of organic development. Professor Owen, whose modesty keeps pace with his fame, will hardly pretend to an infallibility which fails in two such noted instances. Besides, the difficulties which this great anatomist and others have found in sanctioning the development theory, chiefly rest in mistaken assumptions with regard to the constitution of the animal kingdom. It is impossible, as they say, to make out a genealogy in a line of *orders;* but let a fresh naturalist, of equal standing, judge of the theory, after he has considered the animal kingdom in the arrangement now suggested, and I feel assured that its feasibility will receive a more favourable verdict.

The reviewer, however, would not abate one jot of his opinion, although Cuvier, Agassiz, and Owen were all against him! If such be the state of his mind regarding Cuvier, with what face can he condemn St. Hilaire, who only does that towards the dead lion which our critic would also do, supposing the dead lion were equally opposed to his opinion? The grounds for this strong assurance are in personal and immediate observation of facts. "We have examined," says he, "the old records . . . in the spots where nature placed them, and we know their true historical meaning . . . We have visited in succession the tombs and charnel-houses of these old times, and we took with us the clew spun in the fabric of development; but we found this clew no guide through these ancient labyrinths, and, sorely against our will, we were compelled to snap its thread. . . We now dare affirm that geology, not seen through the mist of any theory, but taken as a plain succession of monuments and facts, offers one firm cumulative argument against the hypothesis of development." What first strikes us in this declaration is the tone in which the writer speaks of his own convictions. Cuvier, Agassiz, Owen, may all be wrong; but this writer cannot. He has *seen* what he speaks of. Against "a dogmatical dictation contrary to the sober rules of sound philosophy," (his own words,) there might have surely been some protection in the necessity of retractation to which the best geologists are occasionally reduced. For example, we have Professor Sedgwick, in 1831, undoing a theory he had formerly embraced:

"We now connect the gravel of the plains with the elevation of the newest system of mountains. . . . . That these statements militate against opinions but a few years since held almost universally among us, cannot be denied. But *theories of diluvial gravel, like all other ardent generalizations of an advancing science, must ever be regarded but as shifting hypotheses to be modified by every new fact, till at length they become accordant with all the phenomena of nature.* In retreating, where we have advanced too far, there is neither compromise of dignity nor loss of strength; for in doing this we partake but of the common fortune of every one who enters on a field of investigation like our own."

The contrast between the philosophic modesty of this passage, and the above extract from the Edinburgh reviewer, must be very striking. The reader, who has seen the hollowness of so many of this writer's particular objections to the development theory, can be little at a loss to form an estimate of the personal investigations of which he speaks. He seems to have yet to learn that the necessarily partial investigations which any single geologist may be able personally to make, can give no such amount of the requisite knowledge as may be acquired in another mode of study; that the intellectual powers and preparations of the personal inquirer ought also to be known, before we can set much store even by that light which may be attained by his examinations. It is not uncommon for ordinary mariners to boast of their knowledge of a country from having sailed several times to one of its ports, and for private sentinels to pretend to a superior knowledge of a great battle, in one detachment of which they happened to be engaged. Of such boastings and pretensions I must confess that I am strongly reminded by this writer.

The geological objections to the development theory have now been discussed, and to the public it must be left to decide the question, whether palæontology is favourable or unfavourable to that scheme. I must now advert to the illustrations which the theory derives from physiology, and the objections which have been made to them. The Edinburgh reviewer occupies several of his pages with such objections, but, fortunately, they need not detain us long, as they come to little more than this, that he puts trust in Dr. Clark, of Cambridge, while I have resorted for the support of my general theory to the views advocated by other physiologists.* I may say that these

* Dr. Whewell (preface to *Indications, &c.*) joins the reviewer and others in reprobating the suggestions which have been made in the Vestiges, with regard to a similarity between certain crystallizations, as the figures produced by frost upon windows and the *Arbor Dianæ*, to vegetable forms. The logical merits of the reviewer's mind are here fully indicated, for what does he set

views are presented in my book as correctly as it was possible for me to give them, who am nothing but a general student: in one instance I have employed the language of a popular treatise, (Dr. Lord's)—ridiculed by our reviewer as a book of no authority—merely because the ideas were there presented in a peculiarly intelligible form. The

down as a disproof of these as "traces of secondary means by which the Almighty deviser might establish" the forms of plants? that such crystallizations grow by simple apposition of new matter, and not from germs, as actual vegetables do; the question at issue being merely, whether the electricity concerned in the crystallization might not have some similar effect in determining the forms of the vegetables. I may here remark that I am not alone in surmising some common root for these phenomena. In *Leithead's Electricity*, (1837,) the following passage occurs:—"The form of the route of free electricity is modified by the medium through which it passes, and also by the electric state of such medium, or of that of the relative electrical condition of two bodies between which it is transmitted. If the medium through which it passes possesses a very inferior conducting power, it is obvious that a certain momentum must be requisite to enable the fluid to force its passage to a given distance, and there will be a point at which the momentum of the fluid and the resistance of the body will exactly counterbalance each other; but so soon as the electricity has again accumulated to a sufficient degree to overcome the resistance, it will again force its way in another direction, until it arrives at another point of equilibrium. In this way, we may readily see the *modus operandi* of the electric fluid in imparting regular forms to bodies; and it is highly probable that its action in this respect *extends to the vegetable kingdom, and perhaps operates even on animals*, from the time in which they general aim was, I can honestly declare, to convey the doctrine of the epigenesis of animals, as M. Serres calls it, as an illustration of my subject, considering myself entitled to do so by the position which it has attained in the world. It is, of course, unfortunate for this, as it is for many other doctrines, that it should have an opponent; but this circumstance is fortunately, on the other hand, no adequate ground of condemnation in the judgment

exist in the embryo state . . . Another fact in support of the opinion, that the distinctive forms of bodies are produced by electrical action is, that crystals, and the twigs and leaves of vegetables, all terminate in points or sharp edges, so that the electrical action can proceed no further in increasing the growth, or, in other words, in propelling fresh portions of matter for the extension of the plant, or the crystal, beyond the pointed or edged termination." In a letter of Mr. Crosse to Mr. Leithead, it is stated that, in one of his experiments, there grew, in the inside of an electrified jar filled with hydro-sulphuret of potash, a mineral fungus, three-fourths of an inch in length and one-fourth of an inch in diameter, "*in the shape of a common trumpet-mouthed fungus, which is found on trees.*" "In one experiment," says Mr. Weekes, in a recent letter to myself, "a singularly beautiful electro-vegetation was produced, *a forest in minature*, which, by aid of a good lens, presented many extraordinary appearances, and continued to interest me during many months." It may suit the reviewer and others to scoff at such "resemblances;" but scoffing will not annul, in my mind, the apprehension that there is here some relation of a very interesting kind, the investigation of which may yet give us a deeper insight than we now enjoy into the mysteries of organic being of third parties. I leave, then, the general tenor of this portion of my reviewer's objections, with the remark, that for the one authority which he has called into court, it would be easy to summon many as good on the other side; for instance, Harvey, Grew, Lister, and Meckel. Our critic's own favourite authority—Mr. Owen—would give good evidence: see his *Lectures on the Invertebrated Animals*, where he says that man's embryotic metamorphoses would not be less striking than those of the butterfly, if subjected like them to observation—and then adds, that the human embryo is first vermiform, next stamped with the characters of the apodal fish, afterwards indicative of the enaliosaur, and so forth. There is another most respectable English physiologist—Dr. Roget—who, in his *Bridgewater Treatise*, explicitly says, "that the animals which occupy the highest stations in each series possess, at the commencement of their existence, forms exhibiting a *marked resemblance* to those presented in the permanent condition of the lowest animals of the same series; and that, during the progress of their development, they assume in succession the characters of each tribe, corresponding to their consecutive order in the ascending chain." It is to what has been thus spoken of by such excellent men—what was, I believe, first hinted at by Harvey, and afterwards shadowed forth by John Hunter—that this writer applies the appellation of "a monstrous scheme, from first to last nothing but a pile of wildly gratuitous hypotheses."

This reviewer and others have been eager to point out that "no anatomist has observed the shadow of any change assimilating the nascent embryo to any of the radiata, mollusca, or articulata. Thus are three whole classes [divisions] of the animal kingdom, passed over without any corresponding fœtal type, and in defiance of the law of development." The writer here states what is not true, if any faith is to be placed in one of the first authorities of the age, and one upon which he himself depends; for have we not seen Mr. Owen on the last page affirming that the human embryo is first *vermiform?*—this meaning the form of the worms, a portion of the class Annelides, in one of these lower divisions. That *all* these divisions or sub-kingdoms are not represented in the human embryo is an objection perfectly visionary, for it is not necessary that all should be involved in the ancestry, and therefore analogies to all are not to be looked for. It may be said, then, there is no true difficulty in this quarter.

Perhaps no part of the arguments for the development theory has been more misapprehended, or misrepresented, than this. It is continually said, that the embryo, at any of its particular stages, is not in reality the animal represented by that stage. The Edinburgh reviewer remarks, with regard to the fish stage, "Were the embryo of a mammal thrown off at that time into water (of its own temperature,) it could not support life for a moment." The brain of a child in the seventh month is also said to be not the brain of any of the inferior animals, but a true human brain. The truth is, no one ever pretended that there was such an identity. It is only said that there is a resemblance in general character between the particular embryotic stage of being, and the mature condition and form of the appropriate inferior animal. The particular adaptations, and the character of vital maturity, are all wanting, and therefore it is that the embryo could not live, as the inferior animal represented, if separated from the parent, and really is not that inferior animal.

It may be well, before leaving this part of the subject, to advert to a special charge which this writer, and at least one other,* have brought forward: it is, that I assume, not only that the organic germs of all creatures are alike, but that they are identical. The Edinburgh Review brings a contradiction to this proposition from Dr. Clark. It is wholly unnecessary, for no such assumption was ever made by me. The phrase used in the book was, "Its primary positions [meaning the doctrines of embryonic development] are that the embryos of all animals are not distinguishably different from each other;" which is a very different proposition. In several other instances, propositions are thus misrepresented to afford the glory of a visionary refutation. For example: the idea that there being light in the planets, any inhabitants of these orbs may be presumed to have eyes, as eyes bear a relation to light, is met by him very gravely with the fact, left for him to discover, that animals have eyes before they are born!

I have now reviewed the vestiges of creation, presented in both the geological and physiological records, the former presenting memorials of the actual progression of species, in nearly such a conformity with the general arrangements of the organic kingdoms as we might expect in the pre-

* North American Review, April 1845.

sent state of the science, and the latter affording us proofs—proofs, at least, satisfactory to many of the best anatomists of our age—of a plan of individual development, which may be called the living picture of the advance of species, during the vast ages chronicled by the sedimentary rocks. A third series of vestiges now remains for consideration—namely, those which hint at originations and modifications of organic beings in the current era.

The objections to the occasional production of organic beings, otherwise than *ex ovo*, do not appear to have been softened by the publication of my former volume. All reviewers, with the single exception of the *British and Foreign Medical Review*, have intimated their continued scepticism on this point. The experiment of Professor Schulze, of Berlin, with decaying organic matter floating in a flask to which common air was admitted, after passing through sulphuric acid, thereby being deprived of all animal admixtures—an experiment which ended in the non-production of any animalcules or mould—is pointed to as conclusive. Explanations more or less plausible have also been offered for the origin of the entozoa, the parasites of civilization, the pimelodes cyclopum, etc. I should fear to weary the reader with a new discussion of all these particulars: for the sake of brevity, let me meet the call which the opponents of the development theory usually make, to give it the direct proof which would be afforded by showing one instance, either of the origin of life or the transmutation of species.

The objection of the Edinburgh reviewer, to the alleged transmutation of oats into rye, is that he believes it a fable. This is the opinion of one person, advanced without fact or argument to support it. Let us see, on the other hand, what a greater authority on botanical subjects than he—namely, Dr. Lindley—has stated on the same subject. "At the request," says this learned person, "of the Marquis of Bristol, the Reverend Lord Arthur Hervey, in the year 1843, sowed a handful of oats, treated them in the manner recommended, by continually stopping the flowering stems, and the produce, in 1844, has been for the most part ears of a very slender barley, having much the appearance of rye, with a little wheat, and some oats; samples of which are, by the favour of Lord Bristol, now before us." The learned writer then adverts to the "extraordinary, but certain fact, that in orchidaceous plants, forms just as different as wheat, barley, rye, and oats, have been proved by the most rigorous evidence, to be accidental variations of one common form, brought about no one knows how, but before our eyes, and rendered permanent by equally mysterious agency. Then, says Reason, if they occur in orchidaceous plants, why should they not also occur in corn plants? for it is not likely that such vagaries will be confined to one little group in the vegetable kingdom; it is more rational to believe them to be a part of the *general system* of creation . . . How can we be *sure*, that wheat, rye, oats, and barley, are not all accidental off-sets from some unsuspected species?"* The reader will now be partly able to judge of the value of the unsupported dictum of the reviewer.

There are many other facts that throw a strong light on transmutation, both of plants and animals. So far from there being any decisive proof against this theory, there is no settled conclusion at this moment amongst naturalists, as to what *constitutes a species*. "There is," says Professor Henslow, "*no law whatever hitherto established, by which the limits of variation to a given species can be satisfactorily assigned*, and until some such law be discovered, we cannot expect precision in the details

* Gardeners' Chronicle, August, 1844.

of systematic botany."* "We have agreed," says Bicheno, "that a species shall be that distinct form, orginally so created, and producing, by certain laws of generation, others like itself. There is this inconvenience attending the use of it by naturalists, that it assumes as a fact, that which, in the present state of science, is in many cases a fit subject of inquiry; namely, that species, according to our definition, do exist throughout nature. It is too convenient a term to be dispensed with, even as an assumption; only *care should be taken that we do not accept the abstract term for the fact.*"† Mr. Westwood, speaking of insects, says, "In very extensive genera, the distinctions of species are so minute, that it requires the most practised eye to separate them; and, indeed, there are some groups, the species of which are so intricately blended together, that no two entomologists are agreed as to their distinctness." According to Mr. Haldeman, author of a learned work on the fresh-water mollusks of America, "There are distinct species in that class—among the Unionidæ, for example, [and this is a remark applicable to other departments of the animal

* Magazine of Zoology and Botany, i. 116.

† Linnæan Transactions, xv. 482.

kingdom,] actually differing less from each other than the known varieties of certain variable species, which a Lamarkian might suppose to be of so recent an origin, as not to have yet become settled in the possession of their proper diagnostic characters. Indeed, notwithstanding the assumption to the contrary, by authors who have little practical acquaintance with the details of natural history, the proper discrimination between species and variety, is one of the greatest difficulties which the naturalist has to encounter; and he who is successful in this department is entitled to a rank which comparatively few can attain."*

Of the extent to which modifications may be carried by palpable external conditions, I may now supply a few illustrations. It is well known that fungi and lichens attain to very different appearances in different situations, in conformity with different conditions. Fries, we are told, "asserts that out of the different states of one species (telephora sulphurea,) more than eight distinct genera had been constructed by different authors. It would seem, then, that the absolute number of species among the fungi is not nearly so great as has been usually supposed; and that

* Boston Journal of Natural History.

the kind produced by a decomposing infusion, or a bed of decaying solid matter, will *depend as much upon the influence of the material employed, as upon the germ itself which is the subject of it.*"*

Among the questions proposed by the Academy of Sciences at Haarlem, in 1839, was one upon the following subject—"According to some botanists, Algæ of a very simple structure, placed under favourable circumstances, develop and change into different plants, belonging to genera much more elevated in the scale of organic being; although these same algæ, in the absence of such favourable circumstances, would be fertile, and reproduce their primitive form."† I would ask if this is a point as yet settled in the negative. The original of our cabbage is well known to be a trailing sea-side plant, entirely different from the cabbage in appearance. The cardoon and artichoke are now admitted to be one, and Mr. Darwin was assured by an intelligent farmer that he has seen, in a deserted garden, the latter plant relapsing into the former.

It is well known, that when fresh-water mollusks are exposed for a little time to an influx

* Carpenter's Physiology, p. 62.

† Charlesworth's Magazine of Natural History, ii. 448.

of the sea, those which can survive the change assume considerably different characters. In a fresh-water tertiary formation of the island of Cos, Professor Edward Forbes and Lieutenant Spratt found various fresh-water molluscan shells—paludina, neretina, melanopsis, etc.—which had passed through surprising modifications in the course of three successive groups of deposits, supposed to have been marked by increasing influxes of sea-water. "The lowermost species of each genus were smooth, those of the centre partially plicated, and those of the upper part strongly and regularly ribbed."* This was apparently a retrogression to marine types. The differences in the three cases were greater than those which naturalists usually consider as grounds of specific distinction.

Surely there are here ample evidences of species, or what are usually regarded as such, being variable under changed conditions. It will be said, these changes are all mere variations of specific forms, and the facts do nothing but show that that has been called species which is only variety. But where is this to have its limits? If the cabbage and sea-plant are to be now regarded as one species, it seems to me that we have to go

* Report of Proceedings of the British Association, 1845.—Literary Gazette.

very little further, to come to the lines of successive forms or *stirpes*, which my hypothesis suggests. This view becomes the more striking when we remember that any variations which we now see, take place within a space of time extremely small in comparison with those which geology allows for its phenomena. "Although," says Mr. Haldeman, "we may not be able, artificially, to produce a change beyond a definite point, it would be a hasty inference to suppose that a physical agent acting gradually for ages, could not carry the varia tion a step or two further."

I may here advert to a fallacy which has been one of the principal difficulties in the way of the supposition of every kind of transmutation. It is always taken for granted that the parental animal must be extinguished in consequence of the change. Thus we find a suggestion by M. St. Hilaire that the modern giraffe may be a modification of the sivatherium of the Indian tertiaries, met very complacently by a reference to the discovery of Dr. Falconer, that, in these tertiaries, the giraffe is associated with the sivatherium. So, also, the suggestion that the hare of Siberia, with its curtailed ears, shorter hind legs, and absence of tail, may be a modification of the ordinary hare, has been answered by Professor Owen, with a reference to the fact, that the tailless hare (Lagomys Spelæus) is found as early in the tertiaries as any species of the true genus, Lepus.* Now it is an assumption on the part of those who oppose the transmutation theory, that the original animal shall perish when the new one is produced; and therefore the difficulty is entirely of their own making. The probability is that the modification takes place in an offshoot of the original tribe, which has removed into a different set of circumstances, these circumstances being the cause of the change: thus there is no need to presume that the original tribe is at all affected by any such modification. The case is precisely analogous to that of a colony. We see, for example, the New Englanders change from the original English type, without any necessary effect upon the parent stock. Just so might the giraffe be a changed sivatherium, and yet the sivatherium continue to exist. And in point of fact, there are many animals now living along with their supposed modified descendants. Unless, therefore, it could be proved that the supposed descendant actually preceded in date the animal

* British Fossil Mammalia and Birds, p. 215.

from which it was said to have sprung, objections of this nature can be of no force. The reader will understand that I only adduce the instances of the sivatherium and hare for the sake of illustration, and without undertaking to show that those animals have actually had such modified descendants as may have been attributed to them. I would intreat the candid opponent of the transmutation theory to review the subject in the improved light in which it appears, with this most gratuitous assumption set aside.

With regard to the origination of new life from inorganic elements, the Broomfield experiment would be quite decisive, if any evidence could be admitted for what men are unwilling to believe. The Edinburgh reviewer writes two pages which appear to put the alleged fact much out of countenance; and yet it is true that ridicule, which always proceeds upon assumption, forms their entire composition. He states that specimens of the insect were sent to Paris, where they set a whole conclave of philosophers a-laughing, because they were found to contain ova. It did not occur to him that independent generation is what the development theory presumes of every animal family which may have ever had an origin otherwise than *ex ovo.* Other specimens were sent to London, but there their fate was sealed by their being found to be not a new species, but one then abundant in the country. These circumstances, with a few empty jests, satisfy the critic that there was no independent generation in the case. Against such a conclusion, proceeding upon mere supposition, I adduce careful experiment. During the last three years, Mr. Weekes, of Sandwich, has continued to subject solutions to electric action, and invariably found insects produced in these instances, while they as invariably failed to appear where the electric action was not employed, but every other condition fulfilled. The rigid care taken in these experiments to exclude vitiating circumstances, gives them a high claim to notice, and I therefore present, as an appendix, two letters from Mr. Weekes upon the subject. They cannot fail to be read with interest, and the more so, as they exhibit a man pursuing the investigation of an important natural fact under the most discouraging circumstances. If this new presentment of the Acarus Crossii shall still excite ridicule, I can only regret the mood of mind from which that ridicule arises; but the opposite party must excuse my attaching no importance to anything besides fact and argument. These alleged phenomena are open, like all others, to the test of counter-experiment. Let them be subjected to it in the most rigid manner, and set aside in the case of failure. But to meet them merely with scoffs and jests, or at the most, certain wholly gratuitous assumptions as to a possibly various cause, is not philosophical, and therefore deserves no consideration.

Having thus presented vestiges of laws for the origination and modification of organic being, I must protest against proof of the existence of such laws being held indispensable to the development theory. The earth, we see, has been peopled for ages before man began to observe nature or chronicle his observations. The organic world attained what appears to us completeness, in remote ages. It is a thing done, as individual reproduction is done at the birth of the new creature. We are not, therefore, to expect conspicuous examples of either a new origin of life or a modification of species at the present day. Though, therefore, not one unequivocal instance of such origin and such modification could be presented, it would say nothing positive against the hypothesis that species originated, and made a series of advances in general organization, by the efficacy of law, in times long antecedent to our historical period. We should still have to say that the evidence of such phenomena was to be looked for elsewhere,—namely, in the history of the progress of organic being as chronicled for us by geology, and in the history which physiology affords us of the progress of the individual embryo. Seeing, then, that plants and animals came into existence gradually, in the course of a vast period of time, and in a succession conforming generally to their grades in organization, and the stages through which the embryo of one of the highest has to pass before it attains maturity, we might say that we had seen all that could well be expected in the case, and enough to establish a strong probability for the development theory. Nevertheless, it may be admitted that any evidence of the continued existence of the creative and modifying laws, is still desirable, for the sake of corroboration. And such is the light in which I regard the facts which we possess regarding variations of type, and the production of some of the lower plants and animals by means independent of generation. As in the progress of an individual being, even after birth, we see the laws which preside over reproduction operating still in a faint degree in the defective nutrition which stunts, and the favouring conditions which advance and glorify, the state of infancy and youth, so might we expect that the laws which originally spread the vegetable and animal kingdoms over the earth, would still, perhaps, be traceable as faintly at work, especially in those lower families where life and the modifiable quality are most abundantly imparted. The evidence for the existence of such laws is patent to the exact observation which will give it philosophical certainty, and to such observation I trust it will, in time, be subjected. Meanwhile, I claim its being received as a provisional aid to the theory of development.

Thus closes my review of the objections which have been made to the evidences for an organic creation by law. Such a mode of that creation was, I said at the first, rendered likely by the manifestation of a presidency of law both in the physical arrangements of the universe and in the constitution of our own minds. It seemed to me that, with evidences of law in these things, we had a strong probability established that law had been the mode of the divine working in the whole system revealed to our senses and reason, throughout all ages of its existence. And I believed that we were called upon, not to grasp at every objection to this idea which could be conjured out of the darkness of our imperfect knowledge, as if to save us from a disrelished conclusion, but rather to look with candid minds into nature, and endeavour to discover in what we do know the traces of such an origin of organization as might harmonize with the conceptions forced upon us from other quarters; trusting that there never could be any disadvantage from embracing that view which the balance of reason might show to be the nearest to truth. The question is, to which view does the balance now incline? Whether is it most likely that the Deity produced being and its many-staged theatre in the manner of order or law, or by any different mode of a more arbitrary character; whether, consequently, are we to regard him as ruling the affairs of the world in the manner of an invariable order or otherwise? I say likely—because we are not to expect on any such questions the absolute demonstration which attends a mathematical problem or an unchallengeable writing. We must be content if we only can see a preponderance of reasons for regarding the universe and its Author in one or other of these lights. To be prepared for a decision upon this question, it is proper that the reader should be presented with a sketch of the theory opposed to that of universal order.

When we set about describing this system, we are struck by finding it vague and unsteady, varying with every degree of intelligence in its votaries and every addition made to science. The uneducated man regards the whole system of the world as resulting from, and depending upon, the immediate working and guidance of an almighty being who acts in each case as may seem to him most meet, exactly as human creatures do. Persons of intelligence, again, usually admit a system of general laws, but for the most part entertain it under great reservations, or in connexion with views totally inconsistent with it. We find Dr. Samuel Clark, for instance, admitting a course of nature as the "will of God producing certain effects in a regular and uniform manner," but, this will, "being arbitrary, [an assumption, as far as natural means of knowledge are concerned,] is, he says, as easy to be *altered* at any time as to be *preserved*."

Others cut off particular provinces of nature as exceptions from the plan of constant order.

Whatever part is dubious or obscure, to mankind generally or to themselves in particular, there they rear the torn standard of the arbitrary system of divine rule. Human volitions form such a region to many who know not that Quetelet has reduced these to mathematical formulæ, and that one of our own most popular divines has written a Bridgewater Treatise, to show the predominance of natural law over mind, as a proof of the existence and wisdom of God. Some who give up this domain to law, find footing in other departments of nature upon which science has not as yet poured any clear light. We shall presently see by what weak arguments such exceptions are maintained. Meanwhile, it must be noted as important, that all is uncertainty on this side of the question—a strong presumption, were there no other, against it.

One of the most remarkable reservations made of late years from the system of invariable order is that presented in Dr. Whewell's *History of the Inductive Sciences.* Admitting that nature, as revealed to our senses, is a system of causation, this writer halts when he comes to consider the origin of language and of arts, the origin of species and formation of globes. These he calls palætiological sciences, because, in his opinion, we have to seek for an *ancient and different class of causes*, as affecting them, from any which are now seen operating. "In no palætiological science," says he, "has man been able to arrive at a beginning which is homogeneous with the known course of events. We can, in such sciences, often go very far back, determine many of the remote circumstances of the past series of events, ascend to a point which seems to be near their origin, and limit the hypothesis respecting the origin itself; but philosophers have never demonstrated, and, so far as we can judge, probably never will be able to demonstrate, what was the primitive state of things from which the progressive course of the world took its first departure. In all these paths of research, when we travel far backwards, the aspect of the earlier portions becomes very different from that of the advanced part on which we now stand; but in all cases the path is lost in obscurity as it is traced backwards to its starting point: it becomes not only invisible, but unimaginable; it is not only an interruption, but an abyss which interposes itself between us and any intelligible beginning of things."*

* Philosophy of the Inductive Sciences, *apud* Indications of the Creator.

Here, we have the view of exceptions which is entertained by one of the chief writers of the day, and the superior of one of our greatest academical institutions. The professional position of Dr. Whewell may be held to imply that we should receive from him a view at once leaning to the philosophical, and accommodated as far as possible to the prepossessions expected in a large class of persons. It is remarkable, but not surprising, how weak is the barrier which he has raised to stop our course towards a theory of universal arrangement by ordinary natural law.

The necessity alleged by Dr. Whewell for a different set of causes in the early times of our globe, and with regard to the formation of that globe, is, at the very first, liable to strong suspicion, as reminding us much of that well-known propensity of nations to fill up the first chapters of their history with mythic heroes and giants. The subjects of investigation are remote from common research; they are not, and never could have been, chronicled in the manner of modern facts; we are in the regions of the comparatively unknown—hence, something more magnificent or impressive than ordinary must be supposed. Such is the reasoning, or rather no-reasoning.

The point at which extraordinary causes have to be supposed is evidently quite arbitrary, resting exactly on the limits of the knowledge existing at any time, and always flying further and further back, in proportion as our knowledge increases. Had Dr. Whewell been writing fifty years ago, he would of course have included among his palætiological sciences, the formation of strata, and the intrusions of the granitic and trappean among the aqueous rocks, which ingenuity has since explained by existing causes;—for there is not a single argument for his considering the formation of globes and origin of species as palætiological, which would not have applied with equal force to these phenomena before the days of Pallas and Hutton. Against a theory of mere assumption—a reasoning from ignorance to ignorance—such considerations form serious objections. But let us come to closer argument. Let us inquire how the idea of a different set of causes for the more important of these phenomena, agrees with such exact knowledge as we have attained respecting them.

"According to the nebular hypothesis," says Dr. Whewell, "the formation of this our system of sun, planets, and satellites, was a process of the same kind as those which are still going on in the heavens. . . . But . . the uniformitarian doctrine on this subject rests on most unstable foundations. We have as yet only very vague and imperfect reasonings to show that by such condensation a *material* system such as ours could result; and the introduction of *organized* beings into such a material system is utterly out of the reach of our philosophy. Here . . therefore, we are led to regard the present order of the world as pointing towards an origin altogether of a different kind from anything which our material science can grasp." Because the nebular hypothesis rests on unstable foundations, and "nothing has been pointed out in the existing order of things which has any resemblance or analogy, of any valid kind, to that creative energy which must be exerted in the production of new species,"—*therefore*, according to Dr. Whewell, we are "driven to assume events *not included in the course of nature*," as having formerly taken place. Such is his reasoning. Now let us call to mind a few of the laws ascertained to have been concerned in the cosmical arrangements, leaving for the meantime all that is doubtful in the nebular hypothesis entirely out of view. The proportion of the equatorial to the polar diameter of the earth is exactly what a fluid mass rotating at such a rate of speed would assume any day we might try the experiment. The relative distances of the planets have been determined by the relation of two laws of matter, so thoroughly patent in their working to modern observation, that a mathematician could ascertain this their result and announce it from his closet, although he never had heard of a planetary system in which it was exemplified. There is, surely, here anything but a likelihood that different causes from those now existing and acting, were the immediate means of producing the cosmical arrangements. May we not rather say that, whatever may have been the details of the formation of globes, we possess ample proof that it was a phenomenon evolved by virtue of exactly the same system of order which we see still operating upon earth? As to the origin of organic beings, our knowledge of geology comes to precisely a similar effect. Admitting that we see not now any such fact as the production of new species, we at least know that, while such facts were occurring upon earth, there were associated phenomena in progress, of a character perfectly ordinary. For example, when the earth received its first fishes, sandstone and limestone were forming in the manner exemplified a few years ago in the ingenious experiments of Sir James Hall; basaltic columns rose for the future wonder of man, according to the principle which Dr. Gregory Watt showed in operation before the eyes of our fathers; and hollows in the igneous rocks were filled with crystals, precisely as they could now be by virtue of electric action, as shown within the last few years by Crosse and Becquerel. The seas obeyed the impulse of gentle breezes, and rippled their sandy bottoms as seas of the present day are doing; the trees grew as now by favour of sun and wind, thriving in good seasons and pining in bad; this, while the animals above fishes were yet to be created. The movements of the sea, the meteorological agencies, the disposition which we see in the generality of plants to thrive when heat and moisture were most abundant, were kept up in silent serenity, as matters of simply natural order, throughout the whole of the ages which saw reptiles enter in their various forms upon the sea and land. It was about the time of the first mammals, that the forest of the Dirt Bed was sinking in natural ruin amidst the sea sludge, as forests of the Plantagenets have been doing for several centuries upon the coast of England. In short, *all the common operations of the physical world were going on in their usual simplicity, obeying that order which we still see governing them,* while the supposed extraordinary causes were in requisition for the development of the animal and vegetable kingdoms. There surely hence arises a strong presumption against any such causes. It becomes much more likely that the latter phenomena were evolved in the manner of law also, and that we only dream of extraordinary causes here, as men once dreamt of a special action of deity in every change of wind and the results of each season, merely because they did not know the laws by which the events in question were evolved.

The writer of the critique in the *Edinburgh Review* is another representative of opinion on this subject whose ideas are worthy of notice. These ideas are not very clear, but I shall endeavour to gather them from the various parts of his paper where they are expressed. He says of certain animals (p. 60)—"They were not called into being by any law of nature, but by a power above nature." If he means by a law of nature something independent of the Deity, I entirely concur with him. Most unquestionably, the animals resulted from a power, which is above nature, in the sense of its being the

Author of nature. He adds—"They were created by the hand of God, and adapted to the conditions of the period." If he here means a special exertion of the powers of the Deity, having a regard to special conditions, we part company, for my object is to show that animals were indebted for their gradations of advance to a law generally impressed by the Deity upon matter, and that their external peculiarities are owing immediately to the agency of those very conditions to which they are supposed to have been adapted. I contend that there was no more need for a special exertion to produce (for instance) mammalia, than there is for one to carry a human fœtus on from the sixth to the seventh, or from the eighth to the ninth month. I had remarked in no irreverent spirit, but the contrary, that the supposition of frequent special exertion anthropomorphises the Deity; I find a similar idea expressed by one who will not be suspected of irreverence on such a subject, the pious and amiable Doddridge—"When we assert," says he, "a perpetual divine agency, we readily acknowledge that matters are so contrived as not to need a divine interposition in a different manner from that in which it had been constantly exerted. And it is most evident that an unremitting energy, displayed in such circumstances, *greatly exalts our idea of God, instead of depressing it;* and therefore, by the way, is so much the more likely to be true." The Edinburgh reviewer denies that there is any lowering of the divine character in supposing a system of special exertion. "The law of creation," he says, "is the law of the Divine will, and nothing else besides. . . The fiat of the Almighty was sufficient at all times, and for all the phenomena of the universe, material and moral."

"It may be true," he continues, "that in the conception of the Divine mind there is no difference between the creation of dead matter and its unbending laws, and the creation of organic structures subservient to all the functions of individual life. But such views are, and must be, above our comprehension. . . Each organic structure is a miracle as incomprehensible as the creation of a planetary system; and each structure is a microcosm related to all other worlds within the ken of sense; yet governed by laws and revolving cycles within itself, and implied in the very conditions of its existence. What know we of the God of nature (we speak only of natural means), except through the faculties he has given us, rightly employed on the materials around us? In this we rise to a conception of material inorganic laws, in beautiful harmony and adjustment; and they suggest to us the conception of infinite power and wisdom. In like manner we rise to a conception of organic laws—of means (often almost purely mechanical, as they seem to us, and their organic functions well comprehended) adapted to an end,—and that end only the well-being of a creature endowed with sensation and volition. Thus we rise to a conception both of Divine power and Divine goodness; and we are constrained to believe, not merely that all material law is subordinate to His will, but that he has also (in the way he allows us to see His works) so exhibited the attributes of His will, as to show himself to the mind of man as a personal and superintending God, concentrating his will on every atom of the universe." The reviewer then censures the language used in my book with respect to the idea of special creative efforts. "Does not our author," says he, "see that he binds the divinity (on his dismal material scheme) in chains of fatalism as firmly as the Homeric gods were bound in the imagination of the blind old poet? . . The material system may end in downright atheism;

or, if not, it stops short in the undeviating sequence of second causes. . . Our view, on the contrary, sees, from one end of the scale to the other, the manifestation of a great principle of creation external to matter—of final cause, proved by organic structures created in successive times, and adapted to changing conditions of the earth. It therefore gives us a personal and superintending God who careth for his creatures."

If such be the best view of the opposite theory which a clever scholar and man of science of the present day can give, that theory must certainly be regarded as in a very unpromising condition. He is, we see, for fiats or efforts adapted to special conditions. These may be, in the divine conception, identical with natural laws or the system of order; but we cannot comprehend it. It is not given to our faculties to understand a matter so profound. Immediately after, he informs us that we have only these faculties to look to for information on this very subject; and they tell us—what?—that the world is a system of law! law, however, subordinate to the Divine will. Surely, if our faculties cannot comprehend the point above stated, they must be equally unable to pronounce decisively upon points so abstruse as law being subordinate to will, and the attributes of that will showing us the Deity as a personal and superintending God. Were controversialists entitled thus to assume that the human faculties can pronounce upon one subject in their own way, but are struck powerless on approaching another, tending to an opposite conclusion, there would, of course, be an end of all argument. But even that exercise of the faculties which the reviewer admits of for his own purpose, by no means goes to the conclusion at which he arrives. He refers but to a small portion of the divine works, when he speaks of "organic structures created in successive times and adapted to the changing conditions of the earth." He cannot be permitted to assume that he has proved these to have been produced by special fiats or any other mode of special exertion, "in conformity with changed conditions:" on the contrary, his proposition is *disproved*, for we hear in many instances of conditions suitable for new beings, countless ages before the suitable beings make their appearance, showing that such was not the principle to which we are solely to look for the genesis of animals. But, even though he were more successful on this point, he would still be required to show his theory of fiats, in harmony with a system, the most important facts of which appear, on the contrary, to have taken their present forms and arrangements under the immediate agency of the "Unremitting Energy." As to results which may flow from any particular view which reason may show as the best supported, I must firmly protest against any assumed title in an opponent to pronounce what these are. The first object is to ascertain truth. No truth can be derogatory to the presumed fountain of all truth. The derogation must lie in the erroneous construction which a weak human creature puts upon the truth. And practically it is the true infidel state of mind which prompts apprehension regarding any fact of nature, or any conclusion of sound argument.

The ingenious Agassiz is equally disposed with Dr. Whewell and the Edinburgh Reviewer to except some part of nature as a domain for special intervention; but he wishes the limits of that domain to be rigidly examined, and reprobates the idea that such inquiries are beyond our province. "If," says he, "it is an obligation on science to proclaim the intervention of a divine power in the development of the whole of nature, and if it is to that power alone that we must ascribe all things, it is not the less incumbent on science to ascertain what is the influence which physical forces, left to themselves, exercise in all natural phenomena, and what is the part of direct action which we must attribute to the supreme being, in the revolutions to which nature has been subjected. . . . It is now time for naturalists to occupy themselves likewise, in their domain, in inquiring within what limits we can recognise the traces of a divine interposition, and within what limits the phenomena take place in consequence of a state of things immutably established from the beginning of the creation. Let it not be said that it is not given to man to sound these depths: the knowledge he has acquired of so many hidden mysteries in past ages, promises more extended revelations. It is an error to which the mind, from a natural inclination to indolence, allows itself too easily to incline, to believe impossible what it would take some trouble to investigate. We generally would impose limits to our faculties, rather than increase their range by their exercise; and the history of the sciences is present to tell us, that there are few of the great truths now recognised, which have not been treated as chimerical and blasphemous before they were demonstrated."*

Where men are so much perplexed between two opposite principles, led by science in the one direction and drawn by intellectual indolence or timidity in the other, it is not surprising to find them expressing opinions wholly contradictory. Sir John Herschel some years ago announced views strictly conformable to those subsequently taken of organic creation in my book. "For my part," said he, "I cannot but think it an inadequate conception of the Creator, to assume it as granted that his combinations are exhausted upon any one of the theatres of their former exercise, though, in this, as in all his other works, we are led, by *all analogy*, to suppose that he operates through a series of intermediate causes, and that, in consequence, *the origination of fresh species, could it ever come under our cognizance, would be found to be a natural, in contradistinction to a miraculous process,*—although we perceive no indications of any process actually in progress which is likely to issue in such a result." In his address to the British Association at Cambridge, (1845,)

* Jameson's Journal, 1842.

he said, with respect to my hypothesis of the first step of organic creation—" The transition from an inanimate crystal to a globule capable of such endless organic and intellectual development, is as great a step—as unexplained a one—as unintelligible to us—and in any sense of the word as *miraculous*, as the immediate creation and introduction upon earth of every species and every individual would be!"

The reader will now be able to judge of the views opposed to the theory of universal order. He observes that they are of no distinct unique character, but for the most part follow the measure of ignorance, and are maintained at the expense of consistency. It is not surprising that the idea of an organic creation by special exertion or fiat should be maintained by the advocates of these views, for it is one of the last obscure pieces of scientific ground on which they can show face. One after another, the phenomena of nature, like so many revolted principalities, have fallen under the dominion of order or law; but here is one little province still faithful to the Bœotian government; and as it is nearly the last, no wonder it is so vigorously defended. As, in the political world, however, men do not trust in the endurance of a dynasty which is reduced to a single city or nook of its dominions, so may we expect a speedy extinction to a doctrine which has been driven from every portion of nature but one or two limited fields. Several eminent authors of our age have even pronounced upon the question as already settled. "Our most deeply investigated views of the Divine Government," says the Rev. Dr. Pye Smith, "lead to the conviction that it is exercised in the way of *order*, or what we usually call *law*. God reigns according to immutable principles, that is *by law*, in *every part of his kingdom—the mechanical, the intellectual, and the moral*; and it appears to be most clearly a position arising out of that fact, that *a comprehensive germ which shall necessarily evolve all future developments*, down to the minutest atomic movements, is a more suitable attribution to the Deity, than the idea of a necessity for irregular interferences."*

In *Blackwood's Magazine*, a writer, understood to be a naturalist of distinguished ability, expresses himself in an equally decided manner:—"To reduce to a system the acts of creation, or the development of the several forms of animal life, no more impeaches the authorship of creation,

* Letter to Dr. Carpenter, appendix to Phil. Mag. xvi. (1840).

than to trace the laws by which the world is upheld, and its phenomena perpetually renewed. The presumption naturally rises in the mind, that the same Great Being would adopt the same mode of action in both cases . . . To a mind accustomed, as is every educated mind, to regard the operations of Deity as essentially differing from the limited, sudden, evanescent impulses of a human agent, it is distressing to be compelled to picture to itself, the power of god as put forth *in any other manner than in those slow, mysterious, universal laws, which have so plainly an eternity to work in*; it pains the imagination to be obliged to assimilate those operations, for a moment, to the brief energy of a human will, or the manipulations of a human hand . . . . There are still, indeed, some men of narrow prejudices, who look upon every fresh attempt to reduce the phenomena of nature to general laws, and to limit those occasions on which it is necessary to conceive of a direct and separate interposition of divine power, as a fresh encroachment on the prerogatives of the Deity, or a concealed attack upon his very existence. And yet these very same men are daily appealing to such laws of the creation as have been already established, for their great proofs of the existence and wisdom of God! . . ." He adds, "No, there is nothing atheistic, nothing irreligious, in the attempt to conceive creation, as well as reproduction, carried on by universal laws."*

There is, however, no more interesting or valuable testimony to universal causation than that presented in the system of Logic of Mr. Stuart Mill. If, in the following extract, we were to substitute the creation of organisms for human volitions, it would apply remarkably well to the state of the argument presented in the present volume:

"The conviction that phenomena have invariable laws, and follow with regularity certain antecedent phenomena, was only acquired gradually, and extended itself, as knowledge advanced, from one order of phenomena to another, beginning with those whose laws were most accessible to observation. This progress has not yet attained its ultimate point; there being still one class of phenomena [human volitions], the subjection of which to invariable laws is not yet universally recognised. So long as any doubt hung over this fundamental principle, the various methods of induction which took that principle for granted could

* Review of Vestiges, Blackwood's Magazine, April, 1845.

only afford results which were admissible conditionally; as showing what law the phenomenon under investigation must follow if it followed any fixed law at all. As, however, when the rules of correct induction had been conformed to, the result obtained never failed to be verified by all subsequent experience; every such inductive operation had the effect of extending the acknowledged dominion of general laws, and bringing an additional portion of the experience of mankind to strengthen the evidence of *the universality of the law of causation:* until now at length *we are fully warranted in considering that law*, as applied to all phenomena within the range of human observation, *to stand on an equal footing in respect to evidence with the axioms of geometry itself.*

"I apprehend that the considerations which give, at the present day, to the proof of the law of uniformity of succession as true of all phenomena without exception, this character of completeness and conclusiveness, are the following:—First; that *we now know it directly to be true of by far the greatest number of phenomena:* that there are *none of which we know it not to be true,* the utmost that can be said being, that of some we cannot positively, from direct evidence, affirm its truth; while *pheno-*

menon after phenomenon, as they become better known to us, are constantly passing from the latter class into the former; and in all cases in which that transition has not yet taken place, the absence of direct proof is accounted for by the rarity or the obscurity of the phenomena, our deficient means of observing them, or the logical difficulties arising from the complication of the circumstances in which they occur; insomuch that, notwithstanding as rigid a dependence upon given conditions as exists in the case of any other phenomenon, it was not likely that we should be better acquainted with those conditions than we are. Besides this first class of considerations, there is a second, which still further corroborates the conclusion, and from the recognition of which the complete establishment of the universal law may reasonably be dated. Although there are phenomena, the production and changes of which elude all our attempts to reduce them universally to any ascertained law; yet in every such case, *the phenomenon, or the objects concerned in it, are found in some instances to obey the known laws of nature.* The wind, for example, is the type of uncertainty and caprice, yet we find it in some cases obeying with as much constancy as any phenomena in nature the law of the tendency of fluids to distribute themselves so as to equalize the pressure on every side of each of their particles; as in the case of the trade winds, and the monsoons. Lightning might once have been supposed to obey no laws; but since it has been ascertained to be identical with electricity, we know that the very same phenomenon, in some of its manifestations, is implicitly obedient to the action of fixed causes. *I do not believe that there is now one object or event in all our experience of nature, within the bounds of the solar system at least, which has not either been ascertained by direct observation to follow laws of its own, or been proved to be exactly similar to objects and events, which, in more familiar manifestations, or on a more limited scale, follow strict laws:* our inability to trace the same laws on the larger scale, and in the more recondite instances being accounted for by the number and complication of the modifying causes, or by their inaccessibility to observation."*

The whole question, then, stands thus. For the theory of universal order—that is, order as presiding in both the origin and administration of the world—we have the testimony of a vast number of facts in nature, and this one in addition,—

* System of Logic, ii. 116.

that whatever is reft from the domain of ignorance and made undoubted matter of science, forms a new support to the same doctrine. The opposite view, once predominant, has been shrinking for ages into lesser space, and now maintains a footing only in a few departments of nature which happen to be less liable than others to a clear investigation. The chief of these, if not almost the only one, is the origin of the organic kingdoms. So long as this remains obscure, the supernatural will have a certain hold upon enlightened persons. Should it ever be cleared up in a way that leaves no doubt of a natural origin of plants and animals, there must be a complete revolution in the view which is generally taken of our relation to the Father of our being.

This prepares the way for a few remarks on the present state of opinion with regard to the origin of organic nature. The great difficulty here is the apparent determinateness of species. These forms of life being apparently unchangeable, or at least always showing a tendency to return to the character from which they may have diverged, the idea arises that there can have been no progression from one to another; each must have taken its special form, independently of other forms, directly from the appointment of the Creator. The Edinburgh reviewer says, "they were created by the hand of God and adapted to the conditions of the period." Now, it is, in the first place, not certain that species constantly maintain a fixed character, for we have seen that what were long considered as determinate species have been transmuted into others. Passing, however, from this fact, as it is not generally received among men of science, there remain some great difficulties in connexion with the idea of special creation. First, we should have to suppose, as pointed out in my former volume, a most startling diversity of plan in the divine workings, a great general plan or system of law in the leading events of world-making, and a plan of minute nice operation, and special attention in some of the mere details of the process. The discrepancy between the two conceptions is surely overpowering, when we allow ourselves to see the whole matter in a steady and rational light. There is, also, the striking fact of an ascertained historical progress of plants and animals in the order of their organization; marine and cellular plants and invertebrated animals first, afterwards higher examples of both. In an arbitrary system, we had surely no reason to expect mammals after reptiles; yet in this order they came. The Edinburgh reviewer speaks of the animals as coming in adaptation to conditions; but this is only true in a limited sense. The groves which formed the coal beds might have been a fitting habitation for reptiles, birds, and mammals, as such groves are at the present day; yet we see none of the last of these classes, and hardly any trace of the two first in that period of the earth. Where the iguanodon lived, the elephant might have lived; but there was no elephant at that time. The sea of the Lower Silurian era was capable of supporting fish, but no fish existed. It hence forcibly appears that *theatres of life must have lain unserviceable, or in the possession of a tenantry inferior to what might have enjoyed them for many ages;* there surely would have been no such waste allowed, in a system where Omnipotence was working upon the plan of minute attention to specialties. The fact seems to denote that the actual procedure of the peopling of the earth was one of a natural kind requiring a long space of time for its evolution. In this supposition the long existence of land without land animals, and more particularly, without the noblest classes and orders, is only analogous to the fact, not nearly enough present to the minds of a civilized people, that to this day the bulk of the earth is a waste as far as man is concerned.

Another startling objection is in the infinite local variation of organic forms. Did the vegetable and animal kingdoms consist of a definite number of species adapted to peculiarities of soil and climate, and universally distributed, the fact would be in harmony with the idea of special exertion. But the truth is, that various regions exhibit variations altogether without apparent end or purpose. Professor Henslow enumerates forty-five distinct floras, or sets of plants upon the surface of the earth, notwithstanding that many of these would be equally suitable elsewhere. The animals of different continents are equally various, few species being the same in any two, though the general character may conform. The inference at present drawn from this fact is, that there must have been, to use the language of the Rev. Dr. Pye Smith, "separate and original creations, perhaps at different and respectively distant epochs." It seems hardly conceivable that rational men should give an adherence to such a doctrine, when we think of what it involves. In the single fact that it necessitates a special fiat of the inconceivable Author of this sand-cloud of worlds to *produce the flora of St. Helena,* we read its more than sufficient condemnation. It surely harmonizes far better with our general ideas of nature, to suppose that, just as all else in this far-spread scene was formed by the laws impressed on it at first by its Author, so also was this. An exception presented to us in such a light appears admissible only when we succeed in forbidding our minds to follow out those reasoning processes, to which, by another law of the Almighty, they tend, and for which they are adapted.

I feel that I have dwelt long enough on this part of the question, and yet there are a few geological facts which here call for special comment, and I am loath to overlook them. As is well known, most of the large carnivores and pachyderms of the late tertiary formations very closely resemble existing species; but they are, nevertheless, determined to be distinct species by Professor Owen and other eminent authorities, in consideration of certain peculiarities. The peculiarities, are, in general, trifling, such as differences in the tubercles or groovings of the surface of teeth, or greater or less length of body or extremities;

but no matter of what the differences consist. Enough for the present that they are held by Mr. Owen and his friends to be of that character which are never passed in generation, but necessarily imply a new creation, a separate effort of divine power. Now it so happens that all the tertiary species, or so-called species, have not been changed or extirpated. There is a *Badger* of the Miocene, which cannot be distinguished from the badger of the present day. Our existing *Meles Taxus* is, therefore, acknowledged by Mr. Owen to be "the oldest known species of mammal on the face of the earth." It is in like manner impossible to discover any difference between the present *Wild Cat* and that which lived in the bone caves with the hyæna, rhinoceros, and tiger of the ante-drift era, all of which are said to be extinct species. So also the otter has survived since an early period in the pliocene, while so many larger animals were shifted. The learned anatomist takes occasion from these facts to speak of a survival by small and weak species of geological changes, which have been accompanied by the extirpation of larger and more formidable animals of allied species. The interference from the facts and doctrines of this school is, that Divine Power has seen fit to change the species of elephants, rhinoceroses, tigers, and bears, using special miracles to introduce new ones, one with perhaps an additional tooth, another with a new tubercle or cusp on the third molar, and so forth, while he has seen no occasion for a similar interference with the otter, wild cat, and badger, which accordingly have been left undisturbed in their obscurity. Such may be the belief of men of science, anxious to support a theory; but assuredly it will never be received by any ordinary men of fair understandings who may be able to read and comprehend the works of Mr. Owen. It were too much for even a child's faith. Yet the Edinburgh reviewer, a member of this school, talks of "credulity!"

Perhaps it is but justice to Professor Pictet to notice his partial dissent from the reigning doctrine on this point. This learned person, finding that the elder alluvion of the Swiss valleys presents mammals identical with those which now live there, though accompanied by remains of elephants, and considering further that "the bats, shrews, moles, badgers, hares, &c., of the caverns appear to be identical with our own," concludes that the following was the order of events as they occurred in Europe: "The species now living and some others, were created at the commencement of the diluvial epoch. Partial inundations and changes of temperature caused some of them to perish, such as the mammoth, the species of bear having an arched forehead, the hyænas, the stag with gigantic horns, the rhinoceros, hippopotamus, &c.; but the greater number of the species escaped these causes of destruction, and still live. Besides those which I have mentioned, and others which I have noticed in the body of my work, it is possible, for example, that the *Ursus Priscus may be the original of recent bears*, etc. It may be said," he adds, "that this idea is opposed to the theory of the peculiarity of species in each formation, and to that of successive creations . . . but I cannot, *on that account, refuse to adopt an explanation of facts which seems to me evident.* The state of theoretical palæontology is still too uncertain to allow of our attaching ourselves too strongly to this or that hypothesis. It is the study of facts which is essential, and we must engage in that study unbiassed by preconceived ideas or particular systems."* I would commend this opinion of

* Traité Elémentaire de Paléontologie; i. 359, 1844. Apud Jameson's Journal, Oct. 1845.

one of the first men of science in Europe to those British savans who regard a greater plication of the enamel in a horse's tooth, or a ridge on a turbinated shell, or a spot on a butterfly's wing, as the proof of a special interference of that Deity who wheeled the orbs into space by a tranquil expression of his will. But M. Pictet must himself revise his opinions. He must quickly perceive that the rule which he lays down for there being no new creation since the diluvial epoch is equally conclusive against new creations at any anterior time. There is a persistency of certain shells since the beginning of the tertiaries; if, then, the moles and badgers be, in any degree, a proof that the present bear is a modification of the *Ursus Priscus*, so also are these shells a proof that all the present mammals are modifications of those of the eocene. Several shells, again, of the secondary formation straggling into tertiaries, are not less conclusive, in rigid reasoning, that all the tertiary species were descended from the secondary, although the wide, unrepresented interval at that point, allowed of a greater transition of forms. In short, the whole of the divisions constructed by geologists upon the supposition of extensive introductions of totally new vehicles of life, must give way before the application of this rule, and it must be seen that what they call new species are but variations upon the old. What, then, will remain to be done, before the theory of progressive development be adopted? Only, as the candid reader will readily surmise, that the cultivators of science should allow themselves to follow the dictates of reason, against the behests of prejudices unworthy of them and of their age.

TIME is the true key to difficulties regarding appearances of determinateness in species. Few of us, not even geologists, have ever realized in our minds the extent of time which has elapsed since the beginning of life upon this globe. Mr. Lyell, without intending to favour the development theory, lends us powerful testimony on this point. After showing reason to believe, that about thirty-five thousand years have passed since the Niagara began to cut down the rock through which it flows, during which time the living mollusks, whether marine or terrestrial, are proved to have undergone no change, he thus proceeds—"If such events can take place, while the zoology of the earth remains almost stationary and unaltered, what ages may not be comprehended in those successive tertiary periods, during which the Flora and Fauna of the globe have been almost entirely changed! Yet how subordinate a place in the long calendar of geological chronology do the successive tertiary periods themselves occupy! How much more enormous a duration must we assign to many antecedent revolutions of the earth and its inhabitants! No analogy can be found in the natural world to the immense scale of these divisions of past time, unless we contemplate the celestial spaces, which have been measured by the astronomer. Some of the nearest of these within the limits of the solar system, as, for example, the orbits of the planets, are reckoned by hundreds of millions of miles, which the imagination in vain endeavours to grasp. Yet one of these spaces, such as the diameter of the earth's orbit, is regarded as a mere unit, a mere infinitesimal fraction of the distance which separates our sun from the nearest star. By pursuing still further the same investigations, we learn that there are luminous clouds, scarcely distinguishable by the naked eye, but resolvable by the telescope into clusters of stars, which are so much more remote, that the interval between our sun and Sirius may be but a fraction of this larger distance. *To regions of* *space of this higher order in point of magnitude, we may, probably, compare such an interval of time as that which divides the human epoch from the origin of the coralline limestone, over which the Niagara is precipitated at the Falls.* Many have been the successive revolutions in organic life, and many the vicissitudes in the physical geography of the globe, and often has sea been converted into land, since that rock was formed. The Alps, the Pyrenees, the Himalaya, have not only begun to exist as lofty mountain chains, but the solid materials of which they are composed have been slowly elaborated beneath the sea, within the stupendous interval of ages here alluded to."*

If time, to anything like the amount here insisted on, have really elapsed between the commencement of life and its attaining its highest forms, we must see that the space comprised by the life of an individual, or even that longer portion during which mankind have been watching the wonders of nature, is not sufficient to allow more than a chance of any transition of species being or having been observed, except perhaps in the humble fields where, as was formerly remarked, reproduction is most active and types least defined.

* Travels in North America, i. 52.

If, however, even in our limited command of this grand element, we can detect such transitions as those amongst the cerealia, or in a common infusion, may we not well suppose that much greater have taken place in the course of the vast series of ages here described? Absolute proof on such a point may be impossible; but nearly the same effect may be reached, if we see vestiges of the supposed facts in living phenomena, just as we conclude upon the formation of stratified and igneous rocks from seeing similar phenomena, generally on a smaller scale, taking place before our eyes.

There is another mode of attaining the means of a tolerably definite conclusion, where perfect proof is unattainable. This is to show a portion or fraction of the entire phenomenon, in conformity with the hypothesis as to the whole. Now this can be done in the case under consideration. There are isolated parts of the earth, which we know to have become dry land more recently than others. Such is the Galapagos group of islands, situated in the Pacific, between five and six hundred miles from the American coast. They are wholly of volcanic origin, and are considered by Mr. Darwin as having been raised out of the sea, "within a late geological period." Here, then, is a piece of the world undoubtedly younger, so to speak, than most other portions are in their totality, that is to say, it has been dry land for a much less space of time, though one still considerable. What are the organic productions of this curious archipelago? In the first place, they are "mostly aboriginal creations, found nowhere else," though with an affinity to those of America. Many of them are even peculiar to particular islands in the group. But the remarkable fact bearing on the present inquiry is, that, excepting a rat and a mouse on two of the islands, supposed to have been imported by foreign vessels, *there are no mammals in the Galapagos.* The leading terrestrial animals are reptiles, and these exist in great variety, and in some instances of extraordinary size. Lizards and tortoises particularly abound. There are also birds, eleven kinds of swimmers and waders, and twenty-six purely terrestrial. All this harmonizes with our ideas of the world in general at the time of the oolites. It speaks of *time* being necessary for the completion of the animal series in any scene of its development. The Galapagos have not had the full time required for the completion of the series, and it is incomplete accordingly.* The entire harmony of this fact does, I must confess, strike my mind forcibly. Had there been mammals and no reptiles, it would have been quite different. We should then have said, that one decided fact against the development theory had been ascertained. A minor circumstance in the zoology of these islands is worthy of note. The swimming and wading birds are less diverse from those of the rest of the world than the terrestrial species, all of which, but one,

* In the Vestiges, Australia is spoken of, for the same reason, as apparently a new country, one which has been belated in its physical and organic development. We have there an order, or what is called an order, of mammals, namely, the marsupialia, besides a few monotremata ; all of which may be regarded as only mammalian *apices* of certain bird families. The placental mammalia are wholly wanting. One might suppose that the reasoning on which the comparative recentness of this continent was inferred would have been readily intelligible, and that not even the most ingenious perverseness of opposition could have hung a remark upon it. Yet the Edinburgh reviewer presents a note (p. 58), stating that, on my own scheme of nature, New Holland ought to have been considered as one of the *oldest* countries. "He might have argued (from its flora, its cestraceonts, its trigoniæ, and its marsupials) that it was as old as our oolites ; but this would not have served the good ends of the scheme of development. An amusing example of inconsistency." By old, I presume, is here meant duration in the condition of dry land. I thoroughly agree with the Westminster Review, when it says of this passage, " A more complete miscomprehension of

are decidedly peculiar. The same holds good regarding the shells and the insects. Here we have the terrestrial animals spreading out into numerous variations, according to the greater variety, and the more peculiar character, of the circumstances determining their organization.* Mr. Darwin has likewise observed such facts in the natural history of solitary islands, as induce him to express his belief, that "*the waders, after the innumerable web-footed species, are generally the first colonists of small islands.*" It is his supposition, that the birds in those instances are immigrants;

reasoning we have never met with." Assuredly it may well be held up, as that Review holds it, "as a warning to believers in ex parte criticism." The fact is, since, as Professor Phillips admits, there has been no break in the chain of life from the beginning, our other continents, whatever minor changes they may have undergone, have continued without any entire submergence since at least the commencement of terrestrial life. They are, therefore, older than Australia could be presumed to be, even upon the principle hinted at by the Edinburgh reviewer. But is not that principle utterly absurd, implying as it does that life had stood still in Australia at one point, while it was advancing to the highest forms in other countries? Nay, that the agencies employed in the formation of rocks had been stopped there, for perhaps a third of the time of the earth's existence? The note would not be worthy of this analysis, but that the self-complacency of the writer is so apt to impose upon readers who do not inquire for themselves.

* See Darwin's Journal of a Voyage Round the World, c. xvii.

but I must advert to the fact, as strikingly in harmony with my hypothesis of development, which was certainly formed without any knowledge of this illustration.

Another mode of proof in the difficult circumstances with which we are dealing, is to show that the hypothesis will account, on a principle of law, for certain facts which we must otherwise suppose to be wholly capricious and accidental. The hypothesis is, that, as a general fact, the progress of being in both kinds has been from the sea towards the land. Marine species of plants and animals are supposed to be, in the main, the progenitors of terrestrial species. Life has, as it were, crept out of the sea upon the land. This of course leads us to consider the distribution of vegetable and animal forms in the sea, and the effect which these may have had in determining the Flora or Fauna of particular detached provinces. We would necessarily suppose that any particular Flora or Fauna occupying a certain geographical area in the ocean, would be apt to become the common source of the Flora or Fauna of any masses of land adjoining to it. Now we shall see how the facts harmonize with this view. Wherever there is a group of islands standing much apart, its plants and animals are never found allied to those of any remote region of the earth, but invariably show an affinity to those of the nearest larger masses of land. Thus, for example, the Galapagos exhibit general characters in common with South America; the Cape de Verd islands, with Africa. They are, in Mr. Darwin's happy phrase, satellites to those continents in respect of natural history. Again, when masses of land are only divided from each other by narrow seas, there is usually a community of forms. The European and African shores of the Mediterranean present an example. Our own islands afford another, of far higher value. It appears that the flora of Ireland and Great Britain is various, or rather, that we have five floras, or distinct sets of plants, and that each of these is partaken of by a portion of the opposite continent. There are, 1st, a flora confined to the west of Ireland, and imparted likewise to the north-west of Spain; 2nd, a flora in the southwest promontory of England, and of Ireland, extending across the Channel to the north-west coast of France; 3rd, one common to the south-east of England, and north of France; 4th, an Alpine flora developed in the Scottish and Welsh Highlands, and intimately related to that of the Norwegian Alps; 5th, a flora which prevails over a large part of England and Ireland, "mingling with the other floras, and diminishing, though slightly, as we proceed westward;" this bears intimate relations with the flora of Germany. Facts so remarkable would force the merest fact-collector or species-denominator into generalization. The really ingenious man who lately brought them under notice,* could only surmise, as their explanation, that the spaces now occupied by the intermediate seas must have been dry land at the time when these floras were created. In that case, either the original arrangement of the floras, or the selection of land for submergence, must have been apposite to the case in a degree far from usual. The necessity for a simpler cause is obvious, and it is found in the hypothesis of a spread of terrestrial vegetation from the sea into the lands adjacent. The community of forms in the various regions opposed to each other, merely indicates a distinct marine creation in each of the oceanic areas respectively interposed, and which would naturally advance into the lands nearest to

* See a paper, read by Professor Edward Forbes, at Cambridge, June, 1845, in Literary Gazette, No. 1484.

it as far as circumstances of soil and climate were found agreeable.*

There is still the difficulty of accounting for the origination of the first forms of life in the various lines afterwards pursued to a high development. How was the inorganic converted into the first rudiments of the organic? Whence, and of what nature, was the impulse that first kindled sensation and intelligence upon this sphere? A suggestion on these subjects is hazarded in my book; but though we were to consider the matter as an entire mystery, it is, after all, only so in the same degree, and to the same effect, as the commencement of a new being from a little germ is a mystery to us, although we know that it is one of the most familiar of all natural events. This last marvel we know to be under natural law, though we cannot otherwise explain it. If we can regard the

* It is, perhaps, hardly necessary here to advert to any explanation which might be brought from the diffusion of seeds by ocean currents, because the directness of the opposition of the fields of these floras to each other across the Channel is obviously inconsistent with that idea. In such a case, the constituents of the various floras would have been confused amongst each other by the diversity of currents in the intermediate seas. Mr. Forbes plainly confesses this explanation to be inadmissible in the present case; and, of course, it is not the right explanation in any other.

origin and development of life upon our planet as having been equally under natural law, the whole point is gained; for we are not so much inquiring in order to say *how?* as *was it within or beyond the natural?* We have seen then, as I conceive, that all the associated truths of science go to this point. The whole concur to say, that to believe an exception in this particular of the history of nature, is an absurdity. Difficulties there may be in treating the case positively; some facts of inferior importance may seem to point to an opposite conclusion; but in the balance of the two sets of evidences, those for a universality of natural law downweigh the other beyond calculation.

I have now to allude to a class of objections different from those made on scientific grounds, but fortunately not less easily replied to. It has appeared to various critics, particularly to the writer in the Edinburgh Review, that very sacred principles are threatened by a doctrine of universal law. A natural origin of life, and a natural basis in organization for the operations of the human mind, speak to them of fatalism and materialism. And, strange to say, those, who every day give views of physical cosmogony altogether discrepant in appearance with that of Moses, apply hard names to my book for suggesting a theory of organic creation in the same way liable to inconsiderate odium. I must firmly protest against this mode of meeting speculations regarding nature. The object of my book, whatever may be said of the manner in which it is treated, is purely scientific. The views which I give of this history of organization, stand exactly on the same ground upon which the geological doctrines stood fifty years ago. I am merely endeavouring to read aright another chapter of the mystic book which God has placed under the attention of his creatures. A little liberality of judgment would enable even an opponent of my particular hypothesis, to see that questions as to reverence and irreverence, piety and impiety, are practically determined very much by special impressions upon particular minds. He would see, for example, that the idea of attaching irreverence to a doctrine of natural law is only likely to arise in a mind which has been trained by habit, to regard the divine working as more special in its nature;—precisely as, finding the Edinburgh reviewer speaking of the whole works of the Deity as "vulgar nature" (p. 53), I feel that the impiety which such an idea expresses to my sense, is only impiety to me, who cannot separate nature from God himself, but it is not necessarily so to him, whose education has given him peculiar, and as I think erroneous conceptions on this subject. The absence, however, of all liberality on these points in my reviewers, is striking, and especially so in those whose geological doctrines have exposed them to similar misconstructions. If the men newly emerged from the odium which was thrown upon Newton's theory of the planetary motions, had rushed forward to turn that odium upon the patrons of the dawning science of geology, they would have been prefiguring the conduct of several of my critics, themselves hardly escaped from the rude hands of the narrow-minded, yet eager to join that rabble against a new and equally unfriended stranger, as if such were the best means of purchasing impunity for themselves. I trust that a little time will enable the public to penetrate this policy, and also the real bearing of all such objections. They must soon see that, if a literal interpretation of scripture is an insufficient argument against the true geognostic history of our earth, so also must it be against all associated phenomena, supposing they are presented on good evidence.

"Some persons," says one of my reviewers, "have a vague idea, that there is something derogatory to the lowest form of animal life to have its origin in merely inorganic elements; an idea which results, perhaps, not so much from any subtle and elevated conceptions of life, as from an imagination unawakened to the dignity and the marvel of the inorganic world. What is motion but a sort of life? a life of activity, if not of feeling. Suppose—what, indeed, nowhere exists—an inert matter, and let it be suddenly endowed with motion, so that two particles should fly towards each other from the utmost bounds of the universe; were not this almost as strange a property as that which endows an irritable tissue, or an organ of secretion? Is not the world one—the creature of one God—dividing itself, with constant interchange of parts, into the sentient and the non-sentient, in order, so to speak, to become conscious of itself? Are we to place a great chasm between the sentient and the non-sentient, so that it shall be derogation to a poor worm to have no higher genealogy than the element which is the lightning of heaven, and too much honour to the subtle chemistry of the earth, to be the father of a crawling subject, of some bag, or sack, or imperceptible globule of animal life. No; we have no recoil against this generation of an animalcule by the wonderful chemistry of God; our objection to this doctrine is, that it is not proved."*

As one example of the weakness of the opposition presented by the Edinburgh reviewer on this ground, I may quote a passage in which he has also aimed at convicting me of being enamoured of resemblances, and allowing my senses to be cheated by empty sounds. "Every one," says he, "has heard of the quickness of thought, and who has not heard of the velocity of the galvanic fluid? Therefore, the speed of thought may be reduced to numbers, and a man may think at the rate of 192,000 miles a second! We well know that the author may shelter himself under the juggle of his own words, and tell us that he speaks only of the transmission of our will through the organs of the body. Let him, then, write in more becoming language." Now a man is surely entitled to be judged by his own words, or all judgment might as well cease. After showing that a galvanic battery produces at least some of the effects of the brain, and endeavouring to reconcile ordinary thinkers to the idea of their partial identity by insisting on the almost metaphysical character of the imponderable agents, I said, in a foot-note,

* Blackwood's Magazize, April, 1845.

"If mental action is electric, the proverbial quickness of thought, *that is, the quickness of the transmission of sensation and will*—may be presumed to have been brought to an exact measurement," &c. I leave the reader to judge if language more direct and less illusive than this could have been employed. With regard to the idea conveyed, the critic has perhaps forgot, or never known, that the *merit* of suggesting the identity of the electricity-driven clockwork of Deluc with that operation of the brain which produces the pulsations of the heart, is claimed by his "model of philosophic caution," Sir John Herschel.* The expression used by that philosopher on the occasion, "If the brain be an electric pile," &c., ought, doubtless, to condemn him in the eyes of our critic as a man enamoured of resemblances, and a user of unbecoming phraseology—if our critic be a man of impartiality. But he must (if critics be capable of such weakness) revise his opinion on the subject of resemblances. It might surprise even his self-confident mind to find in what decisive terms their utility as one of the means of advancing in scientific observation is insisted on by this very "model of philosophic caution." He will find the passage at page 94 of the celebrated *Discourse*.

* Discourse on Natural Philosophy, p. 343.

After discussing the whole arguments on both sides in so ample a manner, it may be hardly necessary to advert to the objection arising from the mere fact, that nearly all the scientific men are opposed to the theory of the Vestiges. As this objection, however, is one likely to be of some avail with many minds, it ought not to be entirely passed over. If I did not think there were reasons independent of judgment for the scientific class coming so generally to this conclusion, I might feel the more embarrassed in presenting myself in direct opposition to so many men possessing talents and information. As the case really stands, the ability of this class to give at the present time, a true response upon such a subject, appears extremely challengeable. It is no discredit to them, that they are, almost without exception, engaged, each in his own little department of science, and able to give little or no attention to other parts of that vast field. From year to year, and from age to age, we see them at work, adding no doubt much to the known, and advancing many important interests, but, at the same time, doing little for the establishment of comprehensive views of nature. Experiments in however narrow a walk, facts of whatever minuteness, make reputations in scientific societies; all beyond is regarded with suspicion and distrust. The consequence is, that philosophy, as it exists amongst us, does nothing to raise its votaries above the common ideas of their time. There can, therefore, be nothing more conclusive against our hypothesis in the disfavour of the scientific class, than in that of any other section of educated men. There is even less; for the position of scientific men with regard to the rest of the public is such, that they are rather eager to repudiate, than to embrace general views, seeing how unpopular these usually are. The reader may here be reminded, that there is such a thing in human nature as coming to venerate the prejudices which we are compelled to treat tenderly, because it is felt to be better to be consistent at the sacrifice of even judgment and conscience than to have a war always going on between the cherished and the avowed. Accordingly, in the case of a particular doctrine, which, however unjustly, is regarded as having an obnoxious tendency, it is not surprising that scientific men view it with not less hostility than the common herd. For the very purpose of maintaining their own respect in the concessions they have to make, they naturally wish to find all possible objections to any such theory as that of progressive development, exaggerating every difficulty in its way, rejecting, wherever they can, the evidence in its favour, and extenuating what they cannot reject; in short, taking all the well recognised means which have been so often employed in keeping back advancing truths. If this looks like special pleading, I can only call upon the reader to bring to his remembrance the impressions which have been usually made upon him by the transactions of learned societies and the pursuits of individual men of science. Did he not always feel that, while there were laudable industry and zeal, there was also an intellectual timidity rendering all the results philosophically barren! Perhaps a more lively illustration of their deficiency in the life and soul of Nature-seeking, could not be presented than in the view which Sir John Herschel gives of the uses of science, in a treatise reputed as one of the most philosophical ever produced in our country. These uses, according to the learned knight, are strictly material—it might even be said, sordid—namely, "to show us how to avoid attempting impossibilities—to secure us from important mistakes, in attempting what is, in itself, possible, by means either inadequate, or actually opposed to the end in view—to enable us to accomplish our ends in the easiest, shortest, most economical, and most effectual manner—to induce us to attempt, and enable us to accomplish, objects, which, but for such knowledge, we should never have thought of undertaking."* Such results, it will be felt, may occasionally be of importance in saving a country gentleman from a hopeless mining speculation, or in adding to the powers and profits of an iron-foundry or a cotton-mill; but nothing more. When the awakened and craving mind asks what science can do for us in explaining the great ends of the Author of nature, and our relations to Him, to good and evil, to life and to eternity, the man of science turns to his collection of shells or butterflies, to his electric machine or his retort, and is mute as a child who, sporting on the beach, is asked what lands lie beyond the great ocean which stretches before him. The natural sense of men who do not happen to have taken a taste for the coleoptera or the laws of fluids, revolts at the sterility of such pursuits, and, though fearful of some error on its own part, can hardly help condemning the whole to ridicule. Can we wonder that such, to so great an extent, is their fate in public opinion, when we read the appeal presented in their behalf by the very prince of modern philosophers? Or can we say that

* Discourse on the Study of Natural Philosophy, p. 44.

where such views of "the uses of divine philosophy" are entertained, there could be any right preparation of mind to receive with candour, or treat with justice, a plan of nature like that presented in the Vestiges of Creation? No, it must be before another tribunal, that this new philosophy is to be truly and righteously judged.

It is important that these sentences be not misunderstood. There is both a necessity for the ascertainment of detached facts, that we may attain to the elimination of principles, and a danger in premature generalization, as tending to mislead men from the true road to that result. But, on the other hand, scientific men are seen spending their time in wrong pursuits, merely for want of the tracings which are often supplied for their direction by happy hypotheses. It is to the chilling repression of all saliency in investigation, which characterizes the scientific men of our country and age, that I object, not to a due caution in selecting proper paths in which to venture. The function of hypothesis in suggesting observations and experiments is admitted by one of the most vigorous thinkers of our time. "Without such assumptions, science could never have attained its present state: they are necessary steps in the progress to something more certain. . . . The process of tracing regularity in any complicated and at first sight confused set of appearances, is necessarily tentative: we begin by making any supposition, even a false one, to see what consequences will follow from it; and by observing how these differ from the real phenomena, we learn what corrections to make in our assumption. . . 'Some fact,' says M. Comte, 'is as yet little understood, or some law is unknown: we frame on the subject an hypothesis as accordant as possible with the whole of the data already possessed; and the science, being thus enabled to move forward freely, always ends by leading to new consequences capable of observation, which either confirm or refute, unequivocally, the first supposition.' . . . Let any one watch the manner in which he himself unravels any complicated mass of evidence; let him observe, how, for instance, he elicits the true history of any occurrence from the involved statements of one or of many witnesses: he will find that he does not take all the items of evidence into his mind at once, and attempt to weave them together: the human faculties are not equal to such an undertaking; he extemporizes, from a few of the particulars, a first rude theory of the mode in which the facts took place, and then looks at the other statements one by one, to try whether they can be reconciled with that provisional theory, or what additions or corrections it requires to make it square with them. In this way . . . . we arrive, by means of hypotheses, at conclusions not hypothetical."* It was with the design of thus giving a direction to inquiry, and leading to views of nature previously little thought of, but unspeakably grander than those commonly entertained, that, too eager for truth to regard my own imperfections, I ventured upon my late speculation. When an ordinary reader judges of it, let him remember that the question lies, not between two philosophical theories, but between a theory resting on much scientific evidence and in conformity with all besides of nature which has been ascertained—which may therefore be called *a philosophical theory*—between this, I say, and a *supposition which does not even pretend to have a single scientific fact for its basis,* which on the contrary slights science and sets it aside, seeking for miracle instead of cause,—thus, in a manner, precluding itself from all title to appear before the tribunals of philosophy,—a mere *prejudice,* in short, of the unenlightened intellect, which has nothing but priority in its favour, in which respect

* Mill's System of Logic.

it has no advantage over the notion of the centrality of the earth, or any other of the first impressions of mankind respecting natural phenomena. As a system, moreover, which finds none of the previous labours of science shaped or directed in favour of its elucidation, but all in the contrary way, our theory obviously calls for every reasonable allowance being made for its defects. It may prove a true system, though one half of the illustrations presented by its first explicator should be wrong.

For any mind competent to judge of it by the facts and arguments on which it is founded, there can be little need to insist upon the superiority of the conclusions to which it points, over the results which arise from more limited views of ordinary science. Existing philosophy, halting between the notions of the enlightened and the unenlightened man, leaves us only puzzled. We know not how to regard the phenomena of the world, and our own relation to them. Many sink into a kind of fatalism which paralyzes the faculties; others ascend into fantastic dreams which exercise a not less baleful influence. Some of the disastrous consequences are sufficiently conspicuous; but many more blaze and expend themselves in privacy, known only in the circles where they have been so fatally felt.

The entire conduct of a large portion of society, and more or less that of nearly all the rest, is regulated, or rather cast loose from regulation, by the want of definite ideas regarding that fixed plan of the divine working, on the study and observance of which it is evident that our secular happiness nearly altogether depends. Even acute men of the world are daily seen acting to their own manifest injury, in consequence of their utter ignorance of any system of law pressing around them. With the great bulk of society, life is merely a following of a few inferior instincts, with a perfect blindness to consequences. By individuals and by communities alike, physical and moral evils are patiently endured, which a true knowledge of the system of Providence would cause to be instantly redressed. Daily health and comfort, life itself, are sacrificed through the want of this knowledge. It is not in the heyday of cheerful, active, and prosperous existence, or when we look only to the things which constitute the greatness of nations, that we become sensible of this truth. We must seek for convictions on the subject, beside the death-beds of amiable children, destroyed through ignorance of the rules of health, and hung over by parents who feel that life is nothing to them when these dear beings are no more; in the despairing comfortlessness of the selfish, who have acted through long years on the supposition that the social affections could be starved hurtlessly; in the pestilences ravaging the haunts of poverty, and revenging, in a spreading contagion, the neglect by the rich of the haplessness of their penury and disease-stricken neighbours; in the canker of discontent and crime, which eats into the vitals of a nation in consequence of an unlimited indulgence of acquisitiveness by those possessing the most ready natural resources and standing in the most fortunate positions; in the national degradation and misery which follow wars entered upon in the wantonness of pride, greed, and vanity. Doubtless, were the idea vitally present in the minds of all men, that from laws of unswerving regularity every act, thought, and emotion of theirs helps to determine their own future, both by its direct effects on their fate, and its reflection from the future of their fellow-creatures, and this without any possibility of reprieve or extenuation, we should see society presenting a different aspect from what it does, the sum of human misery vastly diminished, and that of the general happiness as much increased.

I am not to attempt a particular defence of the new view of nature from various odiums thrown upon it, for this can only be rightly done when time has abated prejudice, and shown more clearly the relation of this philosophy to all other views cherished by civilized nations. But I may meanwhile remark its harmony with the great practical principle of Christianity, in establishing the universal brotherhood and social communion of man. And not only this, but it extends the principle of humanity to the meaner creatures also. LIFE is everywhere ONE. The inferior animals are only less advanced types of that form of being perfected in ourselves. Constituted as its head—with a peculiar psychical character and destiny by virtue of that position—we are yet essentially connected with the humbler vehicles of vitality and intelligence, and placed in moral relations towards them. We are bound to respect the rights of animals as of our human associates. We are bound to respect even their feelings. And from obeying these moral laws, we shall reap as certain a harvest of benefit to ourselves, as by obeying any code of law that ever was penned. The rule of force and of cruelty has hitherto prevailed in this department of the world's economy as between man and man; but the day of true knowledge will bring a better rule here also, and the many good qualities of these patient and unresisting ministers of our convenience will yet be acknowledged and dwelt on by all with admiration and love.

Is our own position affected injuriously by this view, or can our relation to the universe and its Author be presumed to be so? Assuredly not. Our character is now seen to be a definite part of a system which is definite. The Deity himself becomes a defined, instead of a capricious being. Power to make and to uphold remains his as before, but is invested with a character of tranquillity altogether new—the highest attribute we can conceive in connexion with power. Viewing him as the author of this vast scheme by the mere force of his will, and yet as the indispensably present sustainer of all; seeing that the whole is constructed upon a plan of benevolence and justice; we expand to loftier, more generous and holy emotions, as we feel that we are essential parts of a system so great and good. The place we hold in comparison is humble beyond all statement of a degree; yet it is a certain and intelligible place. We know where we stand, and have some sense also of our chronological place. The years of our existence occupy a space in that mighty series, during some earlier portion of which this globe, since the theatre of glories and of sorrows numberless, was moulded into form. Arithmetic could state, if we knew it, the connexion between the birth of a babe which saw the light an hour ago, and the time when the elements of our astral system began to resolve themselves into those countless orbs, one of which is Man's, the stage of his long descended history, and the bounds within which all his secular phenomena must ever be confined. The unit of each individuality, great or humble in social regard, takes a fixed place in that march of life which rose unreckoned ages ago, and now goes on to a "weird," which no wizard has pretended to know. We feel that, amidst all the disgrace of trouble and of trespass, we are still the first form of active being after the Greatest, and therefore may well be assured that, immeasurable as is our distance from God, we are still immediately regarded and cared for by him. Surely there is here much to soothe and to encourage. It may be that the individual often suffers innocently to appearance in our present sphere; but then he is part of a system of assured benevolence and justice: having faith in this, he is safe. It may be, as some one has suggested, that there is not only a term of life to the individual, but to the species, and that when the proper time comes, the prolific energy being exhausted, man is transferred to the list of extinct forms. Strange thought, that the beauteous phenomena of personal existence—the thrill of the lover, the mother's smile on cherub infancy, the brightness of loving firesides, the aspirations of generous poets and philosophers, the thought cast up and beyond the earthly, that petard which breaks down every door—the tear of penitence, the meekness of the suffering humble, the ardour of the strong in good causes, all that the great and beneficent of all ages have felt, all that each of us now sees, and muses on, in his home, his people, his age,—that *all these* should be thus resolved; fleeting away whole "equinoxes" into the past, as far as we particular men are concerned, still passing further back as respects the larger personalities called nations, and still further in inconceivable multiplication with regard to the species—gone, lost, hushed in the stillness of a mightier death than has hitherto been thought of! But yet the faith may not be shaken, that that which has been endowed with the power of godlike thought, and allowed to come into communion with its Eternal Author, cannot be truly lost. The vital flame which proceeded from him at first returns to him in our perfected form at last, bearing with it all good and lovely things, and making of all the far-extending Past but one intense Present, glorious and everlasting.

COMMUNICATIONS BY W. H. WEEKES, ESQ.

Referred to at page 120.

DEAR SIR,—Since the details of my first experiments on the production of acari in close atmospheres were given to the world, through the medium of the "Proceedings of the London Electrical Society," *session of* 1842, &c., and, about the same time, circulated among my scientific friends, in a reprint from the above-named work, as stated by you in a foot-note to page 187, first edition of the *Vestiges*, the subject has continued to occupy my attention, while the nature of my researches has been frequently modified by variations in regard to the form of the experiments, and their correlative arrangements.

Incident to the period included by the last three years, many experiments on the subject have been completed; others are even yet in progress; and, however rigid were the conditions in any case adopted, thus much is certain, *that the acari have invariably appeared in the several solutions under electrical influence, while their absence has been as invariably remarked, in spite of the nicest scrutiny, in all negative tests provided to accompany the respective primary experiments.*

The following may be taken as an example of the stringent circumstances under which my latter experiments have been conducted; and although, in my own estimation, the evidence it yields is not one whit more conclusive than the results formerly made known, it is clearly free from certain objections *urged* against the first experiments, and is selected under an impression that, if these conditions fail to show that the electric current is the *agent* by which the laws of organization have been promoted, then we have—maugre the Baconian philosophy—already trusted too much to experimental facts, with a view to the establishment of truth.

It is by no means easy, even if practicable, independent of sketches, to convey a precise idea of the apparatus employed in the experiment I am about to communicate. I will, nevertheless, attempt to describe it with as much brevity and plainness as possible. In the first place, I must mention that the arrangements were originally of a threefold character:—1st, A close vessel containing a saline solution, and above it an artificial atmosphere; 2nd, An open vessel containing the same solution, both acted upon by the same current passing through them from a voltaic battery; 3rd, Two glass jars standing on the same table, as negative tests, and in every way corresponding with the respective primary vessels, excepting that they had no wire appendages, and were unelectrified.

The close vessel consists of a wide-mouthed glass jar, capable of containing a pint and a half of liquid, and is manufactured from the purest and most transparent material. From the top, or shoulder of this jar, ascends, to the height of an inch from the surrounding surface, a remarkably stout and strong neck, which presents an opening of two inches diameter. Into this opening a thick metallic plug or stopper, cast from "fusible alloy," is fitted perfectly air-tight, by a process of long and careful grinding. Perpendicularly through the metallic stopper, and at the distance of an inch from each other, so as to occupy the extremes of an equilateral triangle, are drilled three holes, each rather more than two-tenths of an inch diameter, and into each of these is soldered, air-tight, a corresponding glass tube. The two principal of this series of tubes serve the purpose of insulating a pair of stout copper wires, which pass longitudinally through them, and are united at each end by a joint fusion of the glass and metal. Two other wires of platina proceed from the lower ends of the copper wires to nearly the bottom of the jar, where they terminate in closely-wound spirals, rather more than an inch apart, while the ends of the copper wires, projecting from the upper ends of their respective tubes, have conical cavities drilled out for the reception of a globule of mercury, by means of which communication with the voltaic battery is established. The third tube, passing first to the depth of an inch below the metallic plug, is bent above the latter into a syphon form, and contains in its curvature a globule of mercury weighing about three drachms, which acts as a valve for the occasional escape of gaseous matter generated within the close vessel, and is, at the same time, a guarantee against the ingress of any species of insect life. The mercury employed to form this valve was cautiously distilled from the red sulphuret of that metal.

By the side of the close vessel above described was placed, in the first instance, a glass tumbler, capable of holding half a pint of liquid. Through two pieces of mahogany, cemented to opposite inner surfaces of this second vessel, were made to pass two stout copper wires, terminating, like those adapted to the close jar, in platina spirals a little more than an inch apart near the bottom of the tumbler. The upper ends of these wires were similarly provided with longitudinal cavities also, drilled out for the reception of small globules of mercury, to complete contact and facilitate inter-communication.

On the 2nd of May, 1842, the apparatus, of which a description has been attempted, was set to work after the following manner:—A solution of ferrocyanate of potass, prepared by carefully boiling two ounces of the salt in sixteen ounces of distilled water, being in readiness for the occasion, ten ounces of the liquid were transferred to the glass jar, and immediately after an elastic metal pipe, in communication with an iron bottle in a state of white heat, and from which a stream of pure oxygen rapidly proceeded, was dipped into the solution in the jar. In this way, the gas, without passing through water, or being brought in contact with any external agent, continued to be supplied to the jar, until the entire atmosphere above the solution consisted of oxygen alone, when the metallic plug was deposited instantly in the neck of the jar, so as to cut off all communication with the external air. The open vessel or tumbler being now placed by the side of the close apparatus, and four ounces of the solution before mentioned having been poured into it, the necessary communication between the two vessels was effected by means of suitable wires, and contact at the same time similarly established with the respective poles of a constant battery of ten pairs. By means of this arrangement, the current entered the open vessel first, and then proceeded, through the solution in the close apparatus, in its way to the negative side.

I must here remark that the electric current, immediately on its first application, was observed to decompose the solution with such energy, that I deemed it advisable to suspend the operation until the activity of the battery should be somewhat modified, and it was not until the evening of the 6th of May that I could date the commencement of my experiment.

A circumstantial record of all important changes connected with this experiment has been preserved, up to the present day, embracing a period of three years and three months, but I cannot conclude that any extracts from my memoranda would enhance the interest of the present notice. I shall therefore prefer a brief summary of the results; first premising that two excellent constant batteries have been successively worn out in the undertaking, and that the requisite changes were made without interruption to the electric current, which is now transmitted by a water-battery of twenty pairs, working with the characteristic uniformity of this excellent species of voltaic contrivance. I would further remark that, from the commencement of the experiment, the battery and the respective vessels containing the solutions have been strictly excluded from the light, by means of a screen constructed for the occasion, and the entire proceeding has been confined to a retired room kept constantly locked, no one having access unless accompanied by myself. My general habit has been to visit the arrangement once in two days, for the purpose of noting the progress, supplying the battery with crystals of sulphate of copper, making good the loss of fluids caused by the evaporation, &c.

1. October 19th, 1843—one hundred and sixty-six days from the commencement of the experiment—the first acari seen in connexion therewith, six in number and nearly full-grown, were discovered on the outside of the open glass vessel. On removing two pieces of card which had been laid over the mouth of this vessel, several fine specimens were found inhabiting the under surfaces, and others completely developed and in active motion here and there within the glass.

October 20th.—Making my visit at an hour when a more favourable light entered the room, swarms of acari were found on the cards, about the glass tumbler, both within and without, and also on the platform of the apparatus. At this identical hour Dr. J. Black favoured me with a call, inspected the arrangements, and received six living specimens of the acarus produced from solution in the open vessel. No trace of insect life could at this time be discovered in the close vessel with an oxygen atmosphere. The solution in the open vessel had undergone very slight change of colour, but exhibited a multitude of minute and beautifully coloured crystals with a prevailing tinge of crimson. The solution beneath the oxygen atmosphere, about ten days after the voltaic current began to traverse it, had assumed a reddish-brown appearance, which gradually darkened in colour until scarcely any light could be transmitted through it, or the ascent of gas from either of the electrodes perceived.

2. Myriads of acari continued to be developed from the solution in the open vessel until the 20th of August, 1843, when it was found expedient to determine this division of the experiment, and confine the operation of the electric current solely to the

close arrangement, in which no appearance of insect life had yet been detected. Before removing the open vessel I had, however, the satisfaction to supply therefrom abundance of living specimens to my scientific friends who had kindly interested themselves on the subject, in various parts of England, Scotland, France, and America.

3. In the beginning of the month of June, 1844, rather more than two years from the commencement of these operations, the solution in the close vessel began to manifest signs of a most remarkable change, the results of constant, slow, and almost invisible decomposition. The apparatus was carefully tested, and found, as at first, perfectly air-tight, and the confined liquid was evidently returning to a paler red colour, as well as a partially translucent condition. These latter appearances rapidly increased, and about the beginning of September in the same year, the solution had acquired a light amber colour and perfect transparency, with abundant flakes and scroll-like forms of irregular oxide of iron of a deep orange colour, nearly covering the bottom of the jar. Most of these had, doubtless, been detached in succession from the negative platina spiral, and were conspicuous through the altered solution. It was while engaged in examining this singular accumulation of oxide, by means of an excellent lens, that I saw for the first time an unequivocal proof of the existence of insect life within the close vessel. Several spinous processes of the acari and other remains were detected floating on the surface of the solution, and others attached to the inside of the glass a few lines above the liquid, while, under circumstances somewhat more obscure, several entire dead insects were perceived amidst the flakes resting on the bottom of the jar. An omission—of secondary importance, it is true—was now for the first time apparent in the apparatus: this was the want of a fitting shelf or resting-place for the insects; a circumstance that my kind friend, Andrew Crosse, Esq., when he favoured me with a visit a few weeks after, remarked almost immediately, and said, before he knew that acari had already appeared, "that they

would fall in and be drowned almost as fast as they were produced." Mr. Crosse was right in his conjecture, for although I have latterly watched the proceeding with diurnal care, I have never identified the presence of more than two living insects at the same time within the close apparatus, and these have as speedily as invariably shared the fate of their predecessors. Notwithstanding the omission alluded to, I enjoy an increase of satisfaction in the knowledge that I have kept from my arrangements any substance which by its introduction might have been suspected of vitiating the results, while the main object of the undertaking has in no wise suffered in its accomplishment. I have only to add my belief, founded on considerable experience and much observation, that insect life was first developed in this division of my experiment, sometime in the month of July, 1844, about two years and two months from the commencement.

I am, dear sir, yours faithfully,

W. H. WEEKES.

Sandwich, 2nd Sept. 1845.

To the Author of "Vestiges of the Natural History of Creation."

ELECTRO-VEGETATION.

On the 3rd of October, 1842, I commenced an electro-chemical experiment, which has constantly, since that period, been in progress, and will probably continue for sometime longer. It is not necessary to the present notice that I should detail the objects of this undertaking, as the indications of a successful result induce me to suppose that particulars may eventually be worth communicating to the scientific public. I shall therefore merely state that a cylindrical glass vessel, capable of containing about ten fluid ounces, *with a bottom of porous baked earth,* and open at the top, is suspended in a convenient frame, is about three-fourths filled with a solution of refined sugar in distilled water, receiving occasional supplies, and that the poles of a water-battery of twenty-five pairs terminate within an inch of each other in the solution before mentioned, about an inch also from the bottom of the cylindrical vessel. Through the porous bottom alluded to, the saccharine liquid gradually percolated, during several months—that is, until its minute viaducts became completely obstructed. The solution thus filtered fell into a convenient glazed earthen jar placed under the apparatus, and was occasionally returned to the inside of the glass cylinder.

About the beginning of September, 1843, a small patch of fungus, of a peculiar character, was observed to have commenced forming on the outside of the glass, near its lower rim, but yet not in contact with the line of junction between the glass and its earthen bottom. At this period the solution had ceased to drop through the earthen diaphragm, and the incipient fungus occupied a spot on the outside of the glass *directly opposite the negative electrode* within. This substance having, when first seen, a gelatinous appearance, of a dark-brown colour, by slow degrees extended itself round the lower rim of the glass, forming an irregular band or zone, half an inch in breadth, and throwing out numerous protuberances as it approached the positive side of the arrangement. On the 29th of November, in the same year, the following note relative to this singular production occurs among my memoranda; and as I cannot otherwise better describe its mature appearance, I shall subjoin the extract:—

"The substance of this fungus varies in colour from a light chocolate to that of a dark sanguineous red, and though formerly of a soft texture, it now offers considerable resistance. When viewed with an excellent pocket-lens—the only sort of microscope that can be brought to bear upon it—a most singularly-beautiful species of vegetation is seen to occupy its entire surface, presenting various shades of crimson, green, olive, and green inclining to yellow. In its general appearance it at once suggests the idea of a magnificent forest, consisting of trees and flowering shrubs in miniature. In particular spots, fine, downy, needle-like spires occur in vast multitudes, and these otherwise naked processes rising from the body of the fungus, are surmounted by what appear to be seed-vessels in some instances, and irregular feathery tufts in others." *

This experiment was not designed with any reference to my researches on the development of the electrical acari, but swarms of these creatures appeared incidental to its progress, and, at the time the above note was made, many of them were seen inhabiting the miniature forest on the fungus, where they seemed to thrive amazingly, and to attain a larger size than any I have hitherto seen.

About the autumn of the year 1844, the fungus had extended to the positive side of the arrangement, thus forming a continuous circular band; and it is not the least remarkable feature of its brief history, that immediately on the completion of this event, the luxuriance and beauty of its vegetation were observed rapidly to decline. A portion of the fungous mass still adheres to the glass, but it is no longer an object of special interest.

To what extent this singular and beautiful production is indebted to the action of an electric current constantly, and for a long time, traversing the saccharine liquid, in connexion with which it appeared, I am not prepared, by the assistance of facts, at present to say, but the following suggestions occur to my mind as strong analogical reasons in support of its electrical origin nature, and progress.

1st. I am tolerably conversant with most of the known fungi of this country, but am not acquainted with any species with which the one in question can be identified, or even be said to resemble.

* Shortly after the above note was entered in my memoranda, a small portion of the fungus, with its incumbent vegetation, was submitted to a powerful microscope, and a sketch made in accordance, which for obvious reasons cannot be here introduced.

2nd. The glazed earthen jar placed under the porous bottom of the cylinder to catch the filtered liquid, had, at the time the fungus originated, a considerable quantity of dark saccharine matter resembling concrete molasses therein; this was suffered to remain as a negative test to the electrical character of the fungus, presuming the latter to have had its beginning in a portion of sugary deposit derived from the solution through the porous diaphragm; yet, though the surface of the residuum in the earthen jar presented the usual indications of mouldiness, no appearance of a fungoid kind, or that of minute vegetation, could at any time be detected within the unelectrified jar.

3rd. The commencement of the fungus at a point *precisely corresponding with the negative pole* of the arrangement, its luxuriance and maturity in the intermediate space on the glass cylinder, and its decay on finally reaching the positive side, are in themselves facts pleading strongly in favour of electrical influence over the organization of this remarkable species of vegetation.

W. H. WEEKES.

Sandwich, 5th Sept. 1845.

To the Author of "Vestiges of the Natural History of Creation."

POSTSCRIPT.

SINCE the present edition was put through the press, it is announced that Lord Rosse has discovered the resolvability of the nebula in the sword of Orion, one of those on whose persistency in the cloud-like character under every power of the telescope, the speculations of Sir William Herschel as to the cosmogony were understood to rest. Of course, if this discovery be confirmed, the Herschel speculations must be abandoned.

It becomes important to specify the precise extent to which such a change will affect the general views which I have endeavoured to lay before the public. Let the reader first peruse with attention the portion of this volume between pp. 5 and 25, and the fifth edition of the *Vestiges of the Natural History of Creation*, at pp. 25 and 29, and he will find that the possibility of an abandonment of Herschel's speculations was contemplated, as fraught with no serious damage to the theory of an entire creation in the manner of natural law. In the latter volume, published in January of the present year, the following sentence occurs in Italic print:—"We might, then, entirely dismiss the nebular theory, and still in the relations of the planets, and in the calculations as to their oblate spheroidality, we should have overpowering proof that the cosmical arrangements were produced in the way of natural law"—the conclusion that God acted in this, and no other more arbitrary manner, in his authorship of nature, being the true and sole point which I aimed at establishing.

The fact is, that Herschel and Laplace performed their several parts in our cosmogony independently of each other, and the giving up of the British astronomer's ideas about nebulous matter does not in the slightest degree affect the results attained by the French geometer. These rest on rigid calculations, which can never be gainsaid. What he did, was to prove there being more than four millions of millions of chances against *one*, that the uniform direction of the forty-three motions then ascertained amongst the planetary bodies, was the result of *one primitive cause*, and to show the "dynamical possibility" of the solar system being evolved in its existing forms and arrangements by the ascertained laws of the universe. He afterwards took hold of Herschel's speculations on nebulous matter, to point out the probable *manner* of this evolution; but the Herschelian hypothesis, though provisionally adjuvant, was not absolutely necessary to that of Laplace. The Laplacian cosmogony takes us, independently, to a previous form of matter, different from the present: all that we lose by the abstraction of Herschel's views is, that we see no longer in the sky any presumable specimens of this former state of matter.

Perhaps this is to state the case on its very lowest grounds. When we consider the indications afforded by the original crystalline floor of the earth, and the heat and expansion of its internal materials—when we look to the comets, the Zodiacal Light, and what are called the November Meteors, we can hardly say that we are left by the Parsontown telescope in darkness, as to the previous form of matter. Still, for all the purposes in view, it is sufficient that we see the matter, whatever it was, put into its present form and arrangements in the manner of natural law; and so we unquestionably have seen it, first with our geometrical eyes, in the pages of Laplace, and secondly with our actual vision on the experimental table of Plateau. So, for the meantime, let it rest.

I may take this opportunity of adverting to a preface to the second edition of Dr. Whewell's *Indications of the Creator*, in which the present volume is largely commented on. It appears that the etymology of the term *palætiology*, in connexion with the opinions avowed by Dr. Whewell, has led me to present a different definition, from what he assigns, to those sciences to which he applies the term—namely, the sciences referring to the origin of language and arts, the origin of species, and the formation of globes. He does not call these sciences palætiological, because, "in his opinion, we have to seek for an *ancient and different class of causes*, as affecting them, from any which are now seen operating." Dr. Whewell's actual definition of palætiological sciences was—"those in which the object is to ascend from the present state of things to a more ancient *condition*, from which the present is derived by intelligible causes." The actual extent of this mistake must appear very small, when the reader learns that Dr. Whewell considers the origin of globes, of species, &c., as "events out of the course of nature," which is the point to which my arguments were addressed. It would appear, however, that our opposite views may be more correctly stated in the following manner:—He alleges that science fails to explain to us the events involved in the palætiological sciences. I say that science, read aright, gives us vestiges or traces of the causes of those events, tending to a conviction that they were of the same order as those which at present preside over nature. I am sorry that I cannot compliment the learned Master of Trinity on his generosity, or even fairness, in attributing to me the belief that the essence of the system in which we live, "consists in life growing out of dead matter, the higher animals out of the lower, and man out of brutes," (p. 19.) To establish the independence of the natural world on all but that form of the divine working which we speak of as natural law, it was doubtless necessary to show grounds for believing that species originated in the manner explained in the *Vestiges;* but this belief could never be considered as the essence of the system of order by which God rules the world. If he had not been more eager to use ridicule, or take advantage of popular odium, than to appeal to rigid argument, he might have been checked by a pointed declaration in the present volume, that my object is one "to which the idea of an organic creation by law is only *subordinate and ministrative*," in common with various other doctrines.

It is fully admitted by Dr. Whewell, that the Deity operates by general laws. He is not blind to "the wonderful order and harmony, the gradations and connexions, which run through the forms of animal life." He admits that the organic world has been created according to laws in the Creator's mind, though we do not, he thinks, know those laws. With such beliefs, he at the same time condemns the theory of a natural production of the organic world, as "excluding all supernatural intervention of creative power." Now, if science be obscure, as he says, as to the laws which were in the creative mind, and as to causation generally, one might suppose it to be unable to pronounce, in this decisive manner, either for or against supernatural interferences. Dr. Whewell, however, sees, in the adaptations of organic beings to external circumstances, clear proofs of such supernatural procedure—that, indeed, while the Deity acted according to a plan and fixed laws, the creation of each animal was, nevertheless, a special act on his part. Here it is, therefore, that we have indications of the Creator. Now it appears to me, that Dr. Whewell has here placed the doctrine of a Creator in a really unfavourable light, while the favourable light in which I have placed it is misapprehended by him. He rests his scientific means of belief, negatively, upon our not knowing the mode of the organic creation, and speaks as if the ascertainment of natural procedure were to be fatal to it. Suppose this doctrine were to be received, the discovery of laws establishing natural procedure would tend so far to leave science in an atheistic state, which I have never thought to be necessary. Positively, his means of such belief depend on our seeing a supernatural event in each of the "adaptations" of organic beings. But what if science should come to explain these adaptations upon natural grounds? Here, too, it would inflict a severe blow upon the doctrine which Dr. Whewell seeks to uphold. Let us, on the other hand, consider the theory of a natural origin of species with all their peculiarities in regard to its bearing on the doctrine of a deity. There is a bird called the pique-bœuf, which lives upon larvæ picked out the hides of living cattle, and is found to be enabled to live in this manner by a beak resembling a pair of forceps, and a set of claws allowed to be the most curved of all apart from those of the raptorial birds. According to Dr. Whewell's ideas of design, we must regard this accommodation of bill and claws to a clambering life on the body of another animal, as only to be accounted for by supposing the Deity first to contemplate the existence of cattle with larvæ deposited in their hides, and then to perform the individual act of creating the pique-bœuf, with its peculiar claws and beak. The opposite theory sees in the pique-bœuf only a kind of starling—for it is allied to that genus—which has in the course of time been modified by natural forces in its constitution, to suit a mode of life to which the temptation was placed before it—just such inherent forces as enable one human being to become expert in music, another in reasoning, and so on, though coming to more tangible results. Now I will not here pause upon the comparative merits of the two theories, in regard of their attribution of dignity to the Deity; it is only necessary to remark that the latter view does not necessarily exclude either design, or the Deity, which design is held to imply, for the inherent forces employed in the latter case may have been part of a design, though one of general application, and the wisdom of God may be seen as clearly in the fashion of the pique-bœuf upon the one theory as the other. To such a view of design it seems to me unavoidable that we should come, if we are to look in that direction for proofs of Deity at all, for how can we see rudimentary organs in animals to whom they are useless, and yet maintain that each animal was specially designed and framed? Yet Dr. Whewell has been able to read both the *Vestiges* and the *Explanations* without seeing this.

Dr. Whewell afterwards indulges in the following analogy:— "Let us suppose," says he, "some great sovereign to found a city upon a noble scale, laying it out in streets and markets, and squares and gardens; designing and building halls of justice and temples, palaces and manufactories, shops and private dwellings. Let it be supposed, too, that the founder has in his mind some special style of architecture, so deeply imprinted, that all the edifices which he designs, great and small, public and private, bear traces of this style, and have resemblances in their elements, construction, and decoration. We know that this may be so: that the spirit of connexion and consistency in some architectural styles is so deep and pervasive, that it breaks out in every part. Thus, in a city so built, it is probable that every part would be recognisable by an architectural eye; and after any interval of ages, the skilful antiquary would be able to point out the marks of the all-pervading style, and to show common features in the workshop and the palace—a connexion in masonry between the cottage and the court. But if we suppose a spectator thus able to discover resemblance and connexion in the parts of the city, what should we think of his wisdom, if, on the strength of such resemblances, he were to maintain that all the different kinds of building had, in the history of the city, grown out of some original form of mansion by gradual steps; if he were to hold that the site was first occupied by a few cottages, and that these multiplied, extended, coalesced, retaining in their masonry and structure the traces of their origin, and thus became the great and well-built city? Still more, what should we think him if he were to teach that it was rejecting a 'system of order' in archæology, to believe that the tribunals, and markets, and public walks, and religious edifices, had been originally constructed with a view to their special uses? It appears to me that such a doctrine in archæology would correspond to that 'system of order' in physiology, which makes the higher forms of animal life grow out of the lower. We who reject that system are not blind to the traces of connexion in the various parts of his work, which the great Architect has left; but we cannot, on that account, give up the belief that the foundation of this, our city, was a special act. We can the less abandon this belief, inasmuch as we connect with it the belief that the founder of the city has also given us laws for our conduct, and has not left us to guide ourselves by considering how the city grew up of itself,—if, indeed, there be any means of guidance supplied by such a consideration."

So it appears to this learned person that there is the same

reason to suppose, of two buildings of different sizes and grades of use, but similar architecture, that the larger has grown out of the smaller, as there is to suppose, of certain animals of different grades of organization, but connected in general plan of structure, that the higher have grown out of the lower. He must regard this as a perfect analogy, and very decisive of the argument, or he would not have given it at such length. But what is the difference between the two things? no less than this, that we never see one building grow out of another, or grow at all, in a natural sense, while we do see animals come into existence by natural growing, and, in their embryotic progress, pass through the forms of those beneath them in the scale; it being further known that the animals appeared on earth in a succession broadly conformable to their grades in organization. Such are the dialectics brought to the discussion of my views by one of the most conspicuous general students of science in England. Such are the powers which he has shown to give the most simple apprehension to those views. After such an exposition, can it be reasonably maintained that the scientific mind of England—however creditably industrious, however accurate in particulars—is prepared to give fair judgment upon any great generalizations?

May, 1846.

Printed in the United States
By Bookmasters